Introduction to Environmental Impact Assessment

A Guide to Principles and Practice

Second Edition

Bram F. Noble

OXFORD
UNIVERSITY PRESS

OXFORD
UNIVERSITY PRESS

8 Sampson Mews, Suite 204, Don Mills, Ontario M3C 0H5
www.oupcanada.com

Oxford University Press is a department of the University of Oxford.
It furthers the University's objective of excellence in research, scholarship,
and education by publishing worldwide in

Oxford New York
Auckland Cape Town Dar es Salaam Hong Kong Karachi
Kuala Lumpur Madrid Melbourne Mexico City Nairobi
New Delhi Shanghai Taipei Toronto

With offices in
Argentina Austria Brazil Chile Czech Republic France Greece
Guatemala Hungary Italy Japan Poland Portugal Singapore
South Korea Switzerland Thailand Turkey Ukraine Vietnam

Published in Canada
by Oxford University Press

First Published 2010

Library and Archives Canada Cataloguing in Publication

Noble, Bram F., 1975–
Introduction to environmental impact assessment : a guide to
principles and practice / Bram F. Noble.

Includes bibliographical references and index.
ISBN 978-0-19-542962-6

1. Environmental impact analysis—Textbooks.
2. Environmental impact analysis—Canada—Textbooks. I. Title.

TD194.6.N62 2010 333.71'4 C2009-904331-9

Cover design: Sherill Chapman
Photographs (clockwise from top): Oil Pumpjacks, Alta: Andrew Penner/iStockphoto;
Aerial view of quarry: Yvan Dubé/iStockphoto; Montreal, Que.: Tony Tremblay/iStockphoto;
Off Shore Oil Platforms: Dale Taylor/iStockphoto; Deforestation, Vancouver Island, BC:
Ian Graham/iStockphoto; Car Exhaust: a-wrangler/iStockphoto; McArthur River Mine, Sask.:
Scott Prokop/iStockphoto; Logging Truck: Sally Scott/iStockphoto; Smokestack, Hamilton, Ont.:
Glenn Rogers/iStockphoto; Whiteshell Provincial Park Dam, Man.: Gilles DeCruyenaere/iStockphoto;
Coal plant Silhouette: alohaspirit/iStockphoto; Oil slick: mikadx/iStockphoto.

Printed and bound in Canada.

6 7 8 — 14 13 12

FSC
www.fsc.org
MIX
Paper from
responsible sources
FSC® C004071

Contents

List of Boxes, Tables, and Figures

BOXES

TABLES

FIGURES

Preface: Using This Guide

This book is directed to students of environmental impact assessment (EIA). It is written primarily for introductory courses on the practice of EIA in both academic and professional environments. Current practitioners and administrators may find it particularly useful as a reference book for key EIA principles, design, and concepts.

It is not the intent of this book to explore Canadian or international experiences with EIA application in detail—a text that is sufficiently comprehensive for EIA procedural training should not at the same time attempt to examine the details of the regulatory requirements, experiences, and countless case studies related to impact assessment. That would result in a cumbersome text, and the focus here is on EIA process itself. That said, however, the book does include numerous case studies, primarily from Canadian experience but also drawn from American, European, and Australian contexts.

The main focus of the book is on the fundamental principles and 'good' practices of EIA that are, arguably, common across EIA systems. The intent is that the course instructor will use the book as a framework and integrate case examples particular to the specific EIA system of concern. End-of-unit questions and exercises are designed to facilitate this approach. Since EIA is a subject of interest in many disciplines and to persons of varied backgrounds, the book is interdisciplinary in perspective and attempts to balance discussion on physical and human environments.

The book is organized into five parts, with appendices and a glossary. Each part introduces a number of principles and procedures, along with case study examples and selected methods and techniques. Part I introduces the basic principles of EIA, including the aims and objectives of EIA and an overview of the generic EIA process. Part II examines the nature of environmental impacts and common methods and techniques used to support EIA practice. Part III focuses on EIA principles and procedures, from screening, scoping, impact prediction, and evaluation to post-decision analysis. Part IV introduces two key areas of EIA that are under rapid transition—cumulative effects assessment and strategic environmental assessment—and demonstrates linkages between the two and linkages with project-based EIA. Part V is an EIA postscript, reflecting on the substantive effectiveness of EIA and making a case for a more regional and integrative approach to environmental assessment.

Each of the chapters concludes with a list of key terms, which are defined in the glossary at the end of the book, and study questions and exercises. The two appendices are practical exercises in project and strategic assessment and encompass many of the questions and exercises, as well as the basic principles and procedures of EIA. These exercises are hypothetical but are designed in such a way that they can be tailored to any setting or jurisdiction where the book is studied. They are intended to provide course participants with an opportunity to apply the concepts and procedures demonstrated throughout the book. The exercises may be useful as a term project in academic settings or as a day-long workshop exercise in professional training environments.

Bram F. Noble
Associate Professor of Environmental Assessment
Department of Geography and Planning and School
of Environment and Sustainability
University of Saskatchewan

Acknowledgements

I am grateful to the staff at Oxford University Press for their assistance throughout the manuscript writing and editing process for this second edition and to numerous students and colleagues for their valuable insight and suggestions for improvement over the first edition.

To Deana, Noelle, and Elias

PART I

Environmental Impact Assessment Principles

CHAPTER 1

Aims and Objectives of Environmental Impact Assessment

INTRODUCTION

Environmental, economic, and social changes are inherent to development. While the goal of development is positive change, it can also create potentially adverse impacts. The challenge facing governments and industries is to find ways to support development initiatives while enhancing health and well-being and without adversely affecting the environment. Managing the impacts of development activities on the environment, as well as creating positive environmental outcomes from such activities, is essential if development is to be recognized as sustainable. The interest in environmental issues has been unmatched in recent years, as illustrated by the introduction of new state environmental programs and legislation. With growing populations, technological advancement, and increasing demand for natural resources, the need for common tools and mechanisms to manage the impacts of development effectively is of ever-increasing importance.

DEFINING ENVIRONMENTAL IMPACT ASSESSMENT

Originating in the US **National Environmental Policy Act (NEPA)** of 1969, which became law in 1970, **environmental impact assessment (EIA)** is now among the most widely practised environmental management tools in the world. Initially conceived to ensure that the environmental consequences of major development proposals were considered during decision-making, EIA has been followed by the development of many other forms of impact assessment, including social impact assessment, health impact assessment, strategic environmental assessment, and sustainability assessment, to name a few (Morrison-Saunders and Fischer 2006).

There is no single, universally accepted definition of environmental impact assessment (Box 1.1), and the term is often used interchangeably with 'environmental assessment' (EA) or 'impact assessment' (IA). The International Association for Impact Assessment (IAIA), the leading global authority on the use and best practice of impact assessment, and the UK Institute of Environmental Assessment (IEA) define EIA as:

> The process of identifying, predicting, evaluating, and mitigating the biophysical, social, and other relevant effects of development proposals prior to major decisions being taken and commitments made (IAIA and IEA 1999).

Box 1.1 Definitions of Environmental Impact Assessment

EIA evaluates all relevant environmental and resulting social effects which would result from a project (Battelle 1978).

EIA is an activity designed to identify and predict the impact on human health and well-being of legislative proposals, policies, programs, and operational procedures and to interpret and communicate information about the impacts (Munn 1979).

EIA is the study of the full range of consequences, immediate and long-range, intended and unanticipated, of the introduction of a new technology, project, or program (Rossini and Porter 1983).

EIA is a tool for identifying and predicting the impacts of projects and investigating and proposing means for their management (CEARC 1988).

EIA is a planning tool whose main purpose is to give the environment its due place in the decision-making process by clearly evaluating the environmental consequences of a proposed activity before action is taken (Gilpin 1995).

EIA is a process to predict the environmental effects of proposed initiatives before they are carried out (CEAA 2007).

EIA is often described as an environmental protection tool, a methodology, and a regulatory requirement. Arguably, EIA is all of these, but perhaps most important, it is a 'process' designed to aid decision-making through which concerns about the potential environmental consequences of proposed actions, public or private, are incorporated into decisions regarding those actions. In this regard, EIA can also be viewed as a means of strengthening environmental management processes following the consent decision for development (Morrison-Saunders and Bailey 1999).

EIA is an organized means of gathering information used to identify and understand the effects of proposed projects on the **biophysical environment** (e.g., air, water, land, plants, and animals) as well as on the **human environment** (e.g., culture, health, community sustainability, employment, financial benefits) for people potentially affected. In simplest terms, EIA is an integral component of sound decision-making, serving both an information-gathering and analytical component, used to inform decision-makers concerning the impacts and management of proposed developments. As a formal and frequently regulatory-based process, EIA should not be confused with related environmental studies, such as **environmental site assessments** whose purpose is to determine the nature and extent of contaminant levels at a specific site and to identify cleanup and remediation plans.

EIA SUBSTANTIVE PURPOSES AND OBJECTIVES

The primary purpose of EIA is to facilitate the consideration of environment in planning and decision-making and, ultimately, to make it possible to arrive at decisions and subsequent actions that are more environmentally sustainable. In this regard, EIA should not be seen merely as a mechanism for preventing development that might generate potentially negative environmental effects. If this were the case, then few developments would actually take place! Rather, EIA is intended to:

- systematically identify and predict the impacts of proposed development;
- find ways to avoid or minimize significant negative biophysical and socio-economic impacts;
- identify, enhance, and create potentially positive impacts;
- ensure that development decisions are made in the full knowledge of their environmental consequences.

In many respects, the above commonly held view of EIA can be traced back to the rationalist approach to planning and decision-making that emerged in the 1960s, requiring a technical evaluation as the basis for objective decision-making (Owens, Rayner, and Bina 2004). This rational comprehensive approach is based on the elements of: i) defining the problem; ii) setting goals and objectives; iii) identifying options; iv) assessing the options; v) implementing a preferred solution; and vi) monitoring and evaluation. Core to this approach is the assumption of a well-defined problem characterized by a range of possible options, complete information, and objective decision-makers. Although the rationalist approach has been criticized in recent years (information in EIA is rarely value-free, often incomplete, and frequently constrained or shaped by political factors and societal interests), it remains a valid representation of practice and 'represents the framework within which EIA is often used as a tool for planning and decision-making' (Hanna 2005, 7).

Others, however, disagree and interpret EIA in a very different light—as a process designed to influence government activities and to facilitate public debate about development priorities. In many respects, the perceived objectives of EIA depend on the lens through which it is viewed. A local community resident, for example, may interpret EIA as a public relations tool used by developers and politicians to justify decisions, and non-government organizations may view EIA as a tool to improve stakeholder involvement in development decision-making, whereas a proponent may very well view EIA as an expensive hurdle that must be overcome in order to receive development approval.

Cashmore (2004) argues that it is more relevant to represent the reality of extant EIA as a series of nebulous models, operating along a broad spectrum of philosophies and values. These models concern the role of science in EIA and broadly reflect the range of purposes and objectives of the EIA process (Table 1.1). At one end of this spectrum is the belief that the scientific method provides the basis for EIA theory and practice. In order to be credible, the EIA process must be based on scientific objectives, modelling and experimentation, quantified impact predictions, and hypotheses-testing. At the other end of the spectrum is the belief that EIA is a decision tool used to empower stakeholders and promote an egalitarian society, with a strong green interpretation of sustainability. In this regard, EIA must be deliberative, promote social justice, and help to realize community self-governance.

Table 1.1 Spectrum of EIA Philosophies and Values

← Applied science →	*Analytical science:* EIA serves to inform decisions and enhance scientific understanding. Science in EIA is applied, experimental, and naturalistic.
	Environmental design: EIA serves to inform and influence design decisions. Science in EIA is applied, environmental science for design and engineering.
	Information provision: EIA serves to inform decisions. Science in EIA is largely based on the natural sciences, with limited role for the social sciences.
← Civic science →	*Participation:* EIA is about participatory decision-making. There is an extensive role for both the natural and social sciences.
	Environmental governance: EIA is about deliberative democracy. Science in EIA is largely based on the social sciences, with limited role for the natural sciences.

Source: Based on Cashmore 2004.

The objectives, or substantive purposes, of EIA vary; however, they can be separated into 'output objectives' and 'outcome objectives'. Output objectives are the immediate, short-term objectives of the EIA process and include:

- ensuring that environmental factors are explicitly addressed in decision-making processes concerning proposed developments;
- improving the design of the proposed development;
- anticipating, avoiding, minimizing, and offsetting adverse effects relevant to development proposals on the human and biophysical environment;
- facilitating informed development decision-making.

Outcome objectives are the longer-term objectives that are the products of consistent and rigorous EIA application. Such objectives are realized from a broader environmental and societal perspective and include:

- protecting the productivity and capacity of human and natural systems and ecological functions;
- providing a means for public debate about the nature and direction of development;
- facilitating learning and environmental education;
- facilitating participatory approaches to development and decision-making;
- promoting development that is sustainable.

For example, an immediate output of EIA could be abandonment of environmentally unacceptable actions and mitigation, to the point of acceptability, of the environmental effects of approved proposals. The longer-term outcome of EIA would thus be an overall positive contribution to the environment and society.

In practice, the underlying objectives of EIA are based on three *core values* (Sadler 1996): integrity, sustainability, and utility. Integrity implies that EIA should conform to standards and general principles of 'good practice'. Sustainability suggests that EIA should promote environmentally sound and socially acceptable development. Utility is based on the premise that EIA should provide balanced and credible information for planners and decision-makers to make informed decisions. Together, the desired outcome, according to Sadler, is that an EIA results in information that is accurate and appropriate to the nature of development, identification of likely impacts and desired outcomes, and strategies for environmentally sound management practices. The IAIA and IEA (1999) expanded on these core values in the form of a comprehensive set of *basic principles* for EIA based on the notion of 'best practice' (Figure 1.1). These principles apply to all stages of EIA and are meant to be inter-

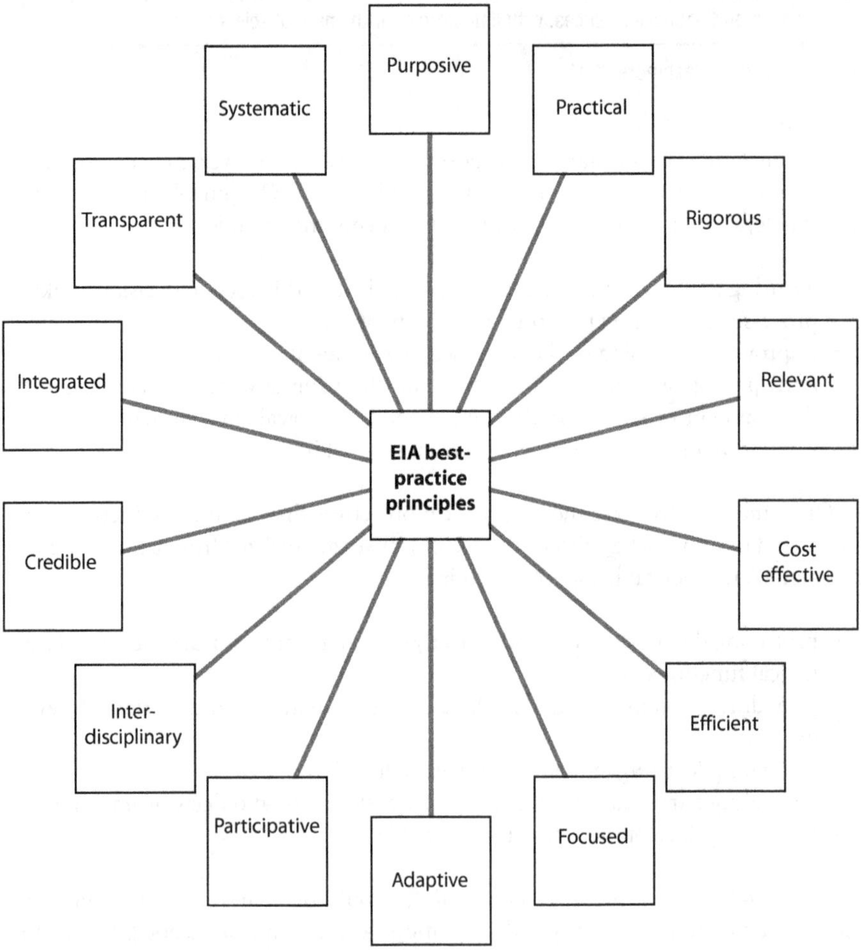

Figure 1.1 Basic principles of 'best practice' EIA.

Source: Based on IAIA and IEA 1999.

preted as a single package, and they are both interdependent and necessary to ensure that EIA fulfills its intended objectives. At the same time, what is 'best practice' in one situation may not necessarily be so in another. 'Best practice' simply means adhering to a number of general principles, or the best way of doing things, as determined by the specific social, political, cultural, and environmental context within which the EIA is taking place.

THE ORIGINS AND DEVELOPMENT OF EIA

EIA Origins

Throughout North America and Western Europe, the 1960s were characterized by a sudden growth in awareness of the relationship between an expanding industrial economy and local environmental change. While many characterize the 1960s as an era of environmental 'idealism', triggered by a number of environmental challenges and sparked by such works as Rachel Carson's *Silent Spring* (1962), the decade did lead to increasing environmental awareness and public demand and pressure on central governments that environmental factors be explicitly considered in development decision-making. As a result of such pressures, the 1960s and early 1970s witnessed the passage of legislation concerning resource protection, hazardous waste management, and control of water and air pollution. However, perhaps the most significant piece of legislation at the time was the National Environmental Policy Act (NEPA) in the United States, which came into effect in January 1970 after its introduction in 1969.

The term 'environmental impact assessment' is actually derived from NEPA, which for the first time required by law that those proposing to undertake certain development projects had to demonstrate that the projects would not adversely affect the environment. To do so, the proponents of a project had to include in their proposal an **environmental impact statement (EIS)** describing the proposed development, the affected environment, likely impacts, and proposed actions to manage or monitor those impacts. During the first 10 years of NEPA, approximately 1,000 EISs were prepared annually. Currently, it is estimated that approximately 30,000 to 50,000 EISs are prepared annually in the US, and some form of EIA system exists in more than 30 individual states.

Prior to the 1970s, development projects were still being assessed, but assessment was limited to technical feasibility studies and, in particular, **cost-benefit analysis.** Cost-benefit analysis (CBA) is an approach to project assessment that expresses impacts in monetary terms. Large-scale development projects such as the 114-metre-high Aswan High Dam in Egypt, for example, constructed during 1960 to 1970 and financed by the US and the United Kingdom, were assessed using CBA techniques. The US Corps of Engineers used CBA for many years to assess and justify large-scale water resource development projects and dam construction in the United States, including the Glen Canyon Dam in Arizona (completed in 1964) and the Oroville Dam in California (completed in 1968). While still commonly used today, obvious drawbacks to CBA include the inability to allocate meaningful dollar values to environmental and human intangibles and the limited scope of fiscal impacts traditionally addressed by CBA. As an 'add-on process' to CBA, EIA was initially intended to incorporate all the potential impacts that were excluded from traditional CBA.

EIA Development

The US NEPA is generally recognized as the pioneer of contemporary environmental assessment. Since its beginnings, EIA has gone through a number of evolutionary phases in the United States and Canada, a pattern repeated to varying degrees throughout the world (see Wood 1995; Sadler 1996).

Initial development. During the early 1970s, immediately following the implementation of NEPA, EIA was characterized by casual and disjointed observations of the biophysical environment, particularly within the local project area. Wider project impacts and potential impact interactions among physical and human environmental components were largely ignored. During this phase, EIA was primarily criticized as a tool used to justify project decisions already made when the process was initiated. The ethos that predominated in this era was 'develop now, minimize associated costs and, if forced to, clean up later' (Barrow 1997). In many cases, as illustrated in Chapter 2, lands were leased and project construction was well underway before an EIA started. The end result was that by the time EIA commenced, many of the opportunities to integrate environmental concerns into project planning and to identify more environmentally sustainable courses of action were foreclosed.

Broadening scope and techniques. By the mid-1970s and up to the early 1980s, EIA efforts became much more highly organized and technically oriented, reflecting the interests of professionals in the natural sciences. Significant attention was devoted to collecting large environmental inventories—i.e., comprehensive descriptions of the biophysical environment in local project areas. The result was impact statements that were thousands of pages in length, often consisting of little more than a compilation of biophysical environmental facts. It is perhaps not surprising then that during this time, project **scoping** (see Chapter 6) was first introduced in an attempt to identify the important issues and data requirements for project assessment. Greater attention turned towards managing adverse project impacts and the risks associated with particular actions, as opposed to creating large environmental inventories. Public review of project proposals and impact assessment processes also emerged as common practice in EIA, and several new innovative impact prediction and assessment techniques were introduced. Perhaps more important, EIA advanced beyond the biophysical environment of the local project in recognition that broader regional and social impacts were also important when evaluating the impacts of proposed project actions on the environment.

Institutional support and integration. The early 1980s to the mid-1990s witnessed rapid growth in EIA, attributed to, among other things, a number of international events, such as the 1987 World Commission on Environment and Development and the 1992 and 1997 Earth Summits, all of which fostered greater international awareness of EIA. During this period, EIA practice turned increasing attention towards physical and social interrelationships associated with project development. **Environment** within the context of EIA was defined as inclusive of not only the biophysical environment but also components of the social and economic environments (e.g., labour markets, demography, housing structure, education, health, values, lifestyles) at multiple spatial and temporal scales (Figure 1.2). Not all current EIA systems, however, adopt such a liberal interpretation of 'environment'. At the Canadian federal level, for example, 'environment' is interpreted as including biophysical sys-

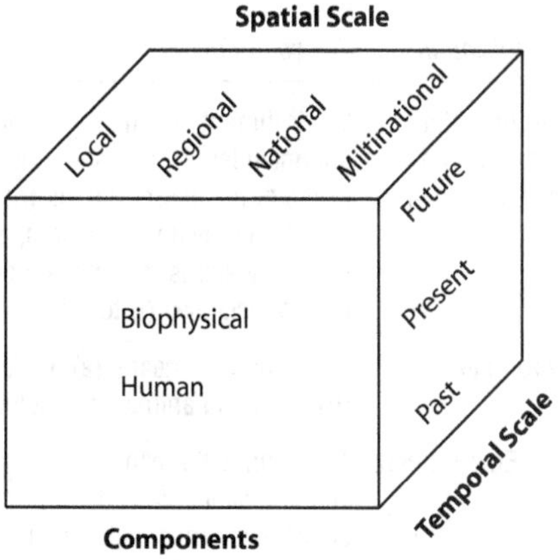

Figure 1.2 Dimensions of 'environment' in EIA.

tems as the primary focus, whereas in many Canadian provincial jurisdictions, 'environment' in EIA is interpreted much more broadly to include social, cultural, and other human dimensions as well (Box 1.2).

EIA emerged in the 1990s as a system-oriented or multi-dimensional approach and involved applications of both qualitative and quantitative models in impact prediction and analysis. Emphasis increasingly turned to the assessment of 'total environmental impacts' and the inclusion of environmental considerations early in the project planning cycle before irreversible decisions were made. This gave way to greater attention to the links between EIA and impact management, social effects assessment, health

Box 1.2 Interpretations of 'Environment' in Canadian EIA

At the Canadian federal level, the primary focus of EIA is on the adverse effects of a project or activity on the biophysical environment. 'Environment', in the context of the Canadian Environmental Assessment Act, means components of the Earth, including land, water, air, organic and inorganic matter, living organisms, and the interacting natural systems of which they are part. While de facto, socio-economic issues are, in their own right, invariably part of the assessment process, the scope of environment has resulted in most of the focus with respect to determining the significance of environmental impacts being concerned with biophysical issues. An 'environmental effect', for example, means, in respect of a project, any change that the project may cause in the (physical) environment and any effect of such change on health or socio-economic conditions, physical and cultural heritage, land and resource use for traditional purposes by Aboriginal persons or on structures, sites, or things of historical, archaeological, palaeontological, or architectural significance. Interpretations of environment under provincial EIA legislation, including that of Newfoundland and Labrador, Ontario, and Saskatchewan, involve what might be considered a broader understanding of environment to encompass direct impacts on communities and on social and cultural systems as well.

continued

Province	*EIA Legislation*	*Scope of Environment*
Alberta	Environmental Protection and Enhancement Act	'Environment' means the components of the Earth and includes: (i) air, land, and water, (ii) all layers of the atmosphere, (iii) all organic and inorganic matter and living organisms, and (iv) the interacting natural systems that include components referred to in subclauses (i) to (iii).
Manitoba	Environment Act	'Environment' means (a) air, land, and water or (b) plant and animal life, including humans.
New Brunswick	Clean Environment	'Environment' means (a) air, water, or soil, (b) plant and animal life, including human life, and (c) the social, economic, cultural, and aesthetic conditions that influence the life of humans or of a community insofar as they are related to the matters described in paragraph (a) or (b).
Newfoundland and Labrador	Environmental Protection Act	'Environment' includes (i) air, land, and water, (ii) plant and animal life, including human life, (iii) the social, economic, recreational, cultural, and aesthetic conditions and factors that influence the life of humans or a community, (iv) a building, structure, machine, or other device or thing made by humans, (v) a solid, liquid, gas, odour, heat, sound, vibration, or radiation resulting directly or indirectly from the activities of humans, or (vi) a part or a combination of those things referred to in subparagraphs (i) to (v) and the interrelationships between two or more of them.
Nova Scotia	Environment Act	'Environment' means the components of the Earth and includes (i) air, land, and water, (ii) the layers of the atmosphere, (iii) organic and inorganic matter and living organisms, (iv) the interacting systems that include components referred to in subclauses (i) to (iii), and (v) for the purpose of Part IV, the socio-economic, environmental health, cultural, and other items referred to in the definition of environmental effect.
Prince Edward Island	Environmental Protection Act	'Environment' includes (i) air, land, and water, (ii) plant and animal, including human, life, any feature, part, component, resources, or element thereof.
Quebec	Environmental Quality Act	'Environment': the water, atmosphere, and soil or a combination of any of them or, generally, the ambient milieu with which living species have dynamic relations.

continued

Province	*EIA Legislation*	*Scope of Environment*
Yukon	Environmental and Socioeconomic Assessment Act	'Environment' means the components of the Earth and includes (a) land, water, and air, including all layers of the atmosphere, (b) all organic and inorganic matter and living organisms, and (c) the interacting natural systems that include components referred to in paragraphs (a) and (b).
Saskatchewan	Environmental Assessment Act	'Environment' means (i) air, land, and water, (ii) plant and animal life, including man, and (iii) the social, economic, and cultural conditions that influence the life of man or a community insofar as they are related to the matters described in subclauses (i) and (ii).

impact assessment, and adaptive environmental management and led to increasing awareness of the need for better monitoring of environmental effects after project implementation. Originally conceived in reaction to growing environmental awareness and to the voice of environmental lobby groups during the late 1960s, EIA was emerging as an increasingly important environmental management tool.

Sustainability initiatives. There is now a growing recognition that EIA should serve as an integrated planning tool for decision-making, characterized by integrating cumulative and global environmental effects, empowering the public, recognizing uncertainties, favouring a precautionary and adaptive approach, and making a positive contribution to sustainability (Gibson 2002). One might say that the breadth of EIA has spread more than its geographical extent. The role of sustainability initiatives, and especially the integration of **sustainability assessment** in EIA, however, has yet to be clearly defined. While some argue that a sustainability focus provides a more comprehensive approach, including socio-economic and biophysical matters and their interrelations and interdependencies as a focal point of EIA, others argue that the natural environment—what EIA was initially designed to protect—loses out in the inherent trade-offs required in such an approach. Kidd and Fischer (2007), for example, assert that the increasing emphasis on integrated assessment may be eroding the focus on the natural environment within the field of EIA as natural environmental concerns are increasingly subordinated to broader sustainability and governance issues and debates.

In that regard, Richard Fuggle, then president of the International Association for Impact Assessment (IAIA), commented in an IAIA newsletter on the disillusionment with measures to promote sustainable development and scepticism about whether EIA is in fact contributing to better decisions (Fuggle 2005). Fuggle suggested that too much might be expected of EIA and that there were perhaps too many different ideas as to what EIA could accomplish. EIA is not a 'magic bullet'—while it complements broader environmental planning, management, and decision-making, it does not replace them. Such disillusionment and scepticism may be due in part to the fact that the fundamental principles of EIA (Figure 1.1) are not being respected. While some Canadian EIAs have been set explicitly within the context of sustainability, such as the Voisey's Bay nickel mine-mill project (Box 1.3), for the most part EIA is still practised, even in the Voisey's Bay case, as a tool for preventing or minimizing environmental problems. In many contexts, EIA has not yet fully advanced beyond this phase, and in many countries EIA is still not conducted prior to the design and approval of projects or plans.

Beyond the individual project. Advancing the sustainability initiative will require increasing the application of EIA principles beyond the project level to address environmental issues at the regional scale and the strategic levels of policy, planning, and program decision-making. This can be accomplished through **strategic environmental assessment** (SEA), the application of environmental assessment principles to policies, plans, and programs. SEA is based on the notion that the benefits of sustainable development trickle down from policy decisions to plans and programs and eventually to individual projects. While SEA is increasingly recognized as a critical tool for advancing the sustainability agenda, considerable process development is still required. The principles and practice of SEA are the focus of discussion in Chapter 13.

Box 1.3 Voisey's Bay Mine-Mill Project

In 1993, a rich nickel-copper-cobalt deposit was discovered at Voisey's Bay in northern coastal Labrador. The deposit has surface dimensions of approximately 800 metres by 350 metres, extends to depths of about 125 metres, and will be mined using open-pit methods. The deposit contains estimated proven and probable mineral reserves of 30 million tonnes, grading 2.85 per cent nickel, 1.68 per cent copper, and 0.14 per cent cobalt. There are an additional estimated 54 million tonnes of mineral wealth at Voisey's Bay, grading 1.53 per cent nickel, 0.70 per cent copper, and 0.09 per cent cobalt, as well as 16 million tonnes of inferred mineral resource, grading 1.6 per cent nickel, 0.8 per cent copper, and 0.1 per cent cobalt.

The project proponent, Voisey's Bay Nickel Company Limited, a subsidiary of Inco Limited, submitted a proposal in 1997 for the development of a mine-mill complex and related infrastructure to produce mineral concentrates at Voisey's Bay. In the absence of Aboriginal land claims agreements, the Voisey's Bay project was subject to review as set out under Canadian federal and Newfoundland provincial environmental assessment processes and pursuant to a memorandum of understanding between the provincial and federal governments, the Labrador Inuit Association, and the Innu Nation.

The public review panel for the Voisey's Bay project issued guidelines for review in which the proponent was required to discuss explicitly the extent to which the project would 'make a positive overall contribution towards the attainment of ecological and community sustainability, both at the local and regional levels' (Voisey's Bay Panel 1997). This was the first major resource development project in Canada for which the impact statement guidelines for the project proponent explicitly identified the sustainability criterion, noting that EIA should go beyond seeking to minimize damage to requiring that an undertaking maximize long-term, durable net gains to the community and the region. Construction of the mine-mill project commenced in 2002. See www.vbnc.com for additional project information.

INTERNATIONAL STATUS OF EIA

Since NEPA, EIA has diffused throughout the world and is currently applied in more than 100 countries (Box 1.4). While NEPA provided the initial basis for EIA, every nation's EIA system is quite distinct, and the enabling legislation, policy directives, and guidelines for EIA vary considerably from one nation to the next. It is not possible to review the status of EIA in all nations, so only a few national EIA systems are highlighted here. It is important to note that while EIA is expanding internationally, it is still relatively new, since many nations have less than 10 years of EIA experience.

Canada

Canada was first to follow the US NEPA beginnings, formally implementing an EIA system in 1973 as a guidelines order through the federal Environmental Assessment and Review Process. It was not until 1995, however, that EIA became entrenched in Canadian law, and subsequent amendments have since taken place, including a 2003

Box 1.4 Selected National EIA Systems

Country	Implemented*	Website to Current System
Armenia	1995	www.mnpiac.am
Australia	1974	www.ea.gov.au
Bulgaria	1991	www.moew.government.bg
Brazil	1981	www.mma.gov.br
Canada	1973	www.ceaa.gc.ca
Chile	1993	www.conama.cl/portal/1255/channel.html
China	1989	www.zhb.gov.cn/english
Colombia	1974	www.minambiente.gov.co
France	1976	www.environnement.gouv.fr
Germany	1985	www.bmu.de
Guyana	1996	www.epaguyana.org
Israel	1982	www.sviva.gov.il
Japan	1984	www.env.go.jp/en/index.html
Madagascar	1995	www.madonline.com/one
New Zealand	1974	www.mfe.govt.nz
Pakistan	1983	www.punjab.gov.pk/epa/index.htm
South Africa	1984	www.environment.gov.za
Thailand	1975	www.thaigov.go.th/index-eng.htm
United States	1969	www.epa.gov

*Many nations have undergone revisions in their EIA systems since first introduced, including moving from guidelines or policy directives to formal legislated systems. Canada, for example, introduced EIA in 1973, but it was not until 1995 that EIA was formally legislated. See also the following website for a directory of environmental agencies of the world: www.inece.org/links_pages/onlineresourcesEnvironmentalagencies.html.

amendment to the Canadian Environmental Assessment Act. Thousands of EIAs are completed each year in Canada, ranging from local initiatives to large-scale resource development projects. Currently, all Canadian provinces have their own EIA systems, with additional EIA systems for Aboriginal land claim areas. At the time this book was written, the Canadian Environmental Assessment Act was under review, and there was speculation that the federal EIA process would be streamlined to significantly reduce the number and types of projects assessed at the federal level. The current Canadian EIA system is discussed in greater detail in Chapter 2.

Australia

Australia was next to formally adopt a system of EIA through its Environmental Protection Act in 1974. The Act was formally implemented in 1975. As with Canada's provinces, most states in Australia have opted for their own form of EIA legislation in which projects of national significance are assessed under a joint state and federal system. In Western Australia, four environmental assessment processes operate under state legislation. The first of these is compulsory for project-level EIA and includes any action that is likely to create significant environmental impacts. Between 40 and 50 of these assessments are completed each year under the guidance of Western Australia's Environmental Protection Agency. A second compulsory process is the referral of land-use plans for environmental assessment by local government and by the state planning agency. This assessment process is seldom triggered—perhaps only two or three times per year; however, hundreds of informal assessments take place whereby local governments and state agencies refer plans for assessment, which are then screened and passed without requiring a full-scale impact assessment. There are also two voluntary processes, one of which is specifically for strategic proposals such as land-use planning strategies, drilling programs, or satellite mining developments. The second concerns the Environmental Protection Agency's authority to report on environmental matters generally. Essentially, proponents outline their future plans and ask for the agency's viewpoint. There is no legally binding outcome, but the process does allow a proponent to pursue an endorsed development option that is more likely to receive approval during formal project impact assessment.

European Union

EIA systems were first introduced in France in 1976, which accounts for by far the most EIAs in Europe, averaging between 5,000 and 6,000 per year (Glasson, Therivel, and Chadwick 1999). It was not until 1985, however, that EIA was formally adopted in Europe through European Union Directive 85/337/EEC, amended in 1997, which set a legal basis for individual member states' EIA regulations (Morris and Therivel 2001). In 1991, the Espoo Convention, signed by the European Commission and several UN Economic Commission for Europe member states, provided a framework for EIA to address potential transboundary impacts. By the mid-1990s, more than 2,500 EISs had been produced in the UK, mostly in England for waste disposal projects, followed by industrial and urban developments, with an average of approximately 300 EISs produced each year (Glasson, Therivel, and Chadwick 1999).

Japan

EIA was discussed in Japan as early as the mid-1970s, and a bill was proposed to parliament in 1976, but it was not until 1984 that EIA was formally adopted through non-mandatory guidelines. Glasson, Therivel, and Chadwick (1999) report that more than half of Japan's local governing authorities have adopted some form of EIA, all of which are more stringently regulated than the national guidelines. A formal legislated EIA system was introduced in Japan in 1997. Currently, about 70 EIAs are prepared in Japan each year.

China

Similarly in China, EIA is implemented through guidelines rather than legislation. A 1979 Environmental Protection Law, amended in 1989, provides the basis for EIA. EIA in China is currently governed by a National Environmental Protection Agency and by more than 2,000 regional and municipal environmental protection bureaus, which receive thousands of EIAs annually.

Developing Nations and Development Agencies

More than 70 developing or transitional nations now have EIA systems in place. The region where perhaps the greatest political transition is currently taking place is Africa, which has EIA systems operating in about a dozen countries, most of which were established within the past decade. Many of these systems operate on an ad hoc basis in response to the requirements of international donor agencies such as the World Bank. The World Bank first introduced EIA requirements in 1989 for evaluating projects it was financing; the Asian Development Bank followed in 1993 with similar requirements (Harrop and Nixon 1999). The Canadian International Development Agency also has EIA requirements for international investment and development projects and recently adopted a system of environmental assessment to address the potential environmental impacts associated with policy and program decisions.

EIA PROCESS

Notwithstanding the diversity of EIA definitions and its adaptations to different international contexts and circumstances, the core elements of EIA are widely agreed upon (Jay et al. 2007). From an applied perspective, EIA can be thought of as a *process* that systematically examines the potential environmental implications of development actions prior to project approval. In short, EIA is simply responsible and proactive planning. The structure of an EIA process, however, is dictated by the specific issues it attempts to address and by the regulatory and legislative requirements within which it operates. While not all EIA systems contain the same elements or specific design procedure, the broad EIA process emanating from the original NEPA and subsequently diffused throughout the world can be thought of as a series of systematic steps (Box 1.5). Although presented here as a linear framework, in practice EIA is an iterative process in which discussions with stakeholders, public review, scoping processes, and post-project evaluations continue to refine impact predictions and management actions. The various stages of the EIA process are explored in greater detail throughout this book.

Several *operating principles* of EIA define how the EIA process should unfold. It should be applied:

- as early as possible in the planning and decision-making stages;
- to all proposals that may generate significant adverse effects or about which there is significant public concern;
- to all biophysical and human factors potentially affected by development, including health, gender, and culture, and cumulative effects;
- consistently with existing policies, plans, and programs and the principles of sustainable development;
- in a manner that allows involvement of affected and interested parties in the decision-making process;
- in accordance with local, regional, national, or international standards and regulatory requirements (IAIA and IEA 1999).

Box 1.5 Generic EIA Process

Project description	Description of the proposed action, including its alternatives, and details sufficient for an assessment.
Screening	Determination of whether the action is subject to an EIA under the regulations or guidelines present, and if so what type or level of assessment is required.
Scoping	Delineation of the key issues and the boundaries to be considered in the assessment, including the baseline conditions and scoping of alternatives.
Impact prediction and evaluation	Prediction of environmental impacts and determination of impact significance.
Impact management	Identification of impact management and mitigation strategies and development of environmental management or protection plans.
Review and decision	Technical and public review of EIS and related documents and subsequent recommendation as to whether the proposed action should proceed and under what conditions.
Implementation and follow-up	Implementation of project and associated management measures; continuous data collection to monitor compliance with conditions and regulations; monitoring the effectiveness of impact management measures and the accuracy of impact predictions.

Public consultation

How the process actually unfolds in practice and the quality of its application, however, vary considerably from one nation to the next (Box 1.6). Box 1.6 illustrates the poor performance of impact monitoring across most national EIA systems. This raises an important question: How does one determine the effectiveness of EIA and the measures put in place to manage impacts in the absence of impact monitoring programs? This issue is discussed in Chapter 10.

BENEFITS OF AND CHALLENGES TO EIA

In addition to managing the impacts of development on the environment, EIA generates a number of direct benefits, including:

- improvements to project planning and design;
- cost savings for proponents through early identification of potentially unforeseen impacts;
- reduction in the role of legality by ensuring early compliance with environmental standards;
- an increase in public acceptability through participation and demonstrated environmental and socio-economic responsibility.

A number of challenges to EIA still must be addressed if EIA is to fulfill its role as a process towards sustainability. Such challenges are reflected in Trudgill's (1990) 'barriers to a better environment', a theoretical framework explaining the common

Box 1.6 Performance of Selected National EIA Systems

Evaluative Component	Australia	Canada	UK	US	New Zealand	Netherlands
Screening	☒	☑	☑	☑	☑	☑
Scoping	☑	☑	☒	☑	☐	☑
Content of EIS	☑	☑	☐	☑	☒	☑
Mitigation	☑	☑	☑	☑	☑	☑
Impact monitoring	☒	☐	☒	☒	☒	☐
Public consultation	☐	☐	☐	☑	☐	☑

☑ Good (criterion met)
☐ Average performance (criterion partially met)
☒ Insufficient (criterion not met)

Source: Based on Petts 1999.

barriers that stand in the way of environmental problem-solving. While Trudgill's work was set primarily within the context of rain forest depletion and acid rain, the framework does capture those barriers to effective EIA, namely:

1. Agreement barrier: Agreement among various interests and practitioners as to what EIA should and can accomplish.
2. Knowledge barrier: Knowledge about the EIA process, its potential value, its role within planning and project decision-making, and knowledge about the actual impacts of projects on the environment post–project implementation.
3. Technology barrier: Availability and appropriate use of methods and techniques to improve the accuracy and precision of project impact prediction and the effectiveness of impact mitigation measures.
4. Economic barrier: The financial cost of doing EIA, including the indirect costs of project development on the environment and society and long-term financial commitment to post-project environmental monitoring.
5. Social barrier: The limited role of public involvement in the EIA process and the limited attention to both direct and indirect social and cultural impacts due to project development, particularly during the monitoring stages.
6. Political barrier: Limited accountability with regard to developers actually doing what they said they would with regard to managing environmental impacts and the use of EIA as a bureaucratic and administrative exercise rather than a process to facilitate informed and responsible decision-making.

These are some of the critical challenges currently faced by EIA. How and whether these challenges are addressed will in large part determine the next phase of EIA evolution and the role of EIA as either a tool for reactive management or one that facilitates the sustainable development of the environment, the economy, and society.

KEY TERMS

biophysical environment
cost-benefit analysis
environment
environmental impact assessment (EIA)
environmental impact statement (EIS)
environmental site assessment
human environment
National Environmental Policy Act (NEPA)
scoping
strategic environmental assessment (SEA)
sustainability assessment

STUDY QUESTIONS AND EXERCISES

1. What are the potential benefits of conducting an environmental impact assessment?
2. How can environmental impact assessment contribute to sustainable development?
3. Obtain a small sample of EISs from your local library or government registry, or access them on-line. Examine the table of contents of each, and create a general list of common elements or issues addressed in each assessment. Are the contents similar in terms of the types of issues and topics addressed?

4. Why was there a sudden interest in environmental impact assessment in North America in the 1960s?
5. Given how environmental impact assessment has evolved over the years, what direction might you expect it to take in the future?
6. What requirements currently exist in your country or province for conducting EIA? When were these requirements implemented, and how have they evolved over time? Is the environment defined to include both physical and human systems? Who is responsible for conducting EIA? How many assessments have been completed to date? Is there a particular sector or type of EIA that dominates? Why might this be so? Do you feel that EIA is meeting its sustainability objectives?
7. Visit some of the websites listed in Box 1.4, and compare the requirements and provisions for EIA to those of your own nation or province.
8. Review some of the challenges to EIA discussed in this chapter. Can you identify one (or more) recent EIA case study in which these challenges were evident? How might they be addressed?
9. As noted in this chapter, EIA has evolved to become one of the most widely applied environmental management tools in the world. Are there certain ethical, cultural, or other elements that should be considered when applying EIA in developed versus developing nations? What elements of the process, if any, might vary from one socio-political or cultural context to the next?
10. Richard Fuggle, former president of the IAIA, raised a question as to whether EIA had passed its 'sell by' date. What do you suppose this means? Do you agree? Provide evidence to support your claim.

REFERENCES

Barrow, C.J. 1997. *Environmental and Social Impact Assessment: An Introduction*. London: Arnold.

Battelle, C. 1978. *Environmental Evaluation in Project Planning*. Columbus, OH: Battelle Environmental Laboratories.

Carson, R. 1962. *Silent Spring*. Boston: Houghton Mifflin.

Cashmore, M. 2004. 'The role of science in environmental impact assessment: Process and procedure versus purpose in the development of theory'. *Environmental Impact Assessment Review* 24: 403–26.

CEAA (Canadian Environmental Assessment Agency). 2007. 'Basics of environmental assessment'. www.ceaa.gc.ca.

CEARC (Canadian Environmental Assessment Research Council). 1988. *The Assessment of Cumulative Effects: A Research Prospectus*. Ottawa: Supply and Services Canada.

Fuggle, R. 2005. 'Have impact assessments passed their "sell by" date?' *Newsletter of the International Association for Impact Assessment* 16 (3): 1, 6.

Gibson, R.B. 2002. 'From Wreck Cove to Voisey's Bay: The evolution of federal environmental assessment in Canada'. *Impact Assessment and Project Appraisal* 20 (3): 151–9.

Gilpin, A. 1995. *Environmental Impact Assessment: Cutting Edge for the 21st Century*. Cambridge: Cambridge University Press.

Glasson, J., R. Therivel, and A. Chadwick. 1999. *Introduction to Environmental Impact Assessment: Principles and Procedures, Process, Practice and Prospects*. 2nd edn. London: University College London Press.

Hanna, K., Ed. 2005. *Environmental Impact Assessment: Practice and Participation*. Don Mills, ON: Oxford University Press.

Harrop, D.O., and A.J. Nixon. 1999. *Environmental Assessment in Practice*. Routledge Environmental Management Series. London: Routledge.

IAIA and IEA. 1999. *Principles of Environmental Impact Assessment Best Practice*. Fargo, ND: IAIA.

Jay, S., et al. 2007. 'Environmental impact assessment: Retrospect and prospect'. *Environmental Impact Assessment Review* 27: 287–300.

Kidd, S., and T. Fischer. 2007. 'Towards sustainability: Is integrated appraisal a step in the right direction?' *Environment and Planning C—Government and Policy* 25 (2): 233–49.

Morris, P., and R. Therivel, Eds. 2001. *Methods of Environmental Impact Assessment*. 2nd edn. London: Taylor and Francis Group.

Morrison-Saunders, A., and J. Bailey. 1999. 'Exploring the EIA/environmental management relationship'. *Environmental Management* 24 (3): 281–95.

Morrison-Saunders, A., and T. Fischer. 2006. 'What is wrong with EIA and SEA anyway? A sceptic's perspective on sustainability assessment'. *Journal of Environmental Assessment Policy and Management* 8 (1): 19–39.

Munn, R.E. 1979. *Environmental Impact Assessment: Principles and Procedures*. New York: Wiley.

Owens, S., T. Rayner, and O. Bina. 2004. 'New agendas for appraisal: Reflections on theory, practice, and research'. *Environment and Planning A* 36: 1,943–59.

Rossini, F.A., and A.L. Porter. 1983. *Integrated Impact Assessment*. Social Impact Assessment Series. New York: Perseus Books.

Sadler, B. 1996. *Environmental Assessment in a Changing World: Evaluating Practice to Improve Performance*. Final report of the International Study of the Effectiveness of Environmental Assessment. Fargo, ND: IAIA.

Trudgill, S. 1990. *Barriers to a Better Environment*. London: Bellhaven Press.

Voisey's Bay Panel. 1997. *Environmental Impact Statement (EIS) Guidelines for the Review of the Voisey's Bay Mining and Mill Undertaking*. Voisey's Bay Mine and Mill Environmental Assessment Panel.

Wood, C. 1995. *Environmental Impact Assessment: A Comparative Review*. London: Longman Scientific and Technical.

CHAPTER 2

A Brief Overview of Environmental Impact Assessment in Canada

EIA IN CANADA

Canada is recognized internationally as a nation that has contributed significantly to the development and advancement of EIA policy and practice. Since its inception in the early 1970s, thousands of EIAs have been completed in Canada. In 2006–7, for example, 8,789 EIAs were either in progress or completed at the federal level alone (Table 2.1). Most of them (99 per cent) were for relatively small-scale, routine development projects and undertakings—referred to as screening assessments—with others classified as comprehensive study or review panel assessments for larger, more significant undertakings (see Chapter 5 for a discussion of the various types of EIA in Canada). This chapter presents a brief overview of Canadian EIA systems. It is not the intent here to discuss Canadian EIA practice, procedure, and requirements in detail, since these elements are addressed throughout the text and covered in greater depth in Hanna 2009. Rather, the chapter briefly outlines Canadian EIA under federal, provincial, and territorial governments and land claims agreements and illustrates the legal development of EIA in Canada.

CANADIAN EIA SYSTEMS

Environmental impact assessment in Canada is enshrined in the law of the provinces, territories, Aboriginal governments, and the federal government (Figure 2.1). Since Canada is a confederation, EIA on the federal level is not binding on the provinces and territories. Rather, responsibility for EIA is divided among the provinces and territories and the federal government, with laws, regulations, EIA objectives, and procedures varying considerably from one jurisdictional system to the next (Bitter 2008). At the time of Confederation, for example, the four 'founding provinces' received authority to make laws governing their natural resources. The Constitution Act, 1930, transferred further responsibility for the development of natural resources, and hence their management, from the Dominion of Canada to the various other provincial governments of the day, including Manitoba, Bitish Columbia, Alberta, and Saskatchewan. The Constitutional Act, 1982, amended the original Constitution to give all provinces the authority to make laws and decisions concerning the development and management of their natural resources.

Table 2.1 Statistical Summary of EIA Activity under the Canadian Environmental Assessment Act

			EIA Decisions 2006–7			*Follow-up Programs*		
EIA Type	*Ongoing 2006–7*	*Initiated 2006–7*	*Not likely to cause significant adverse effects*	*Likely to cause significant adverse effects*	*Terminated*	*Ongoing 2006–7*	*Follow-up program initiated 2006–7*	*Follow-up program completed 2006–7*
Screening	3,192	5,507	4,397	0	330	145	85	64
Comprehensive study	37	23	8	0	1	11	3	0
Review panel	24	6	2	0	0	1	1	0

Source: Compiled based on the Canadian Environmental Assessment Agency's 'Statistical summary' for the 2006–7 fiscal year. www.ceaa.gc.ca/010/index_e.htm.

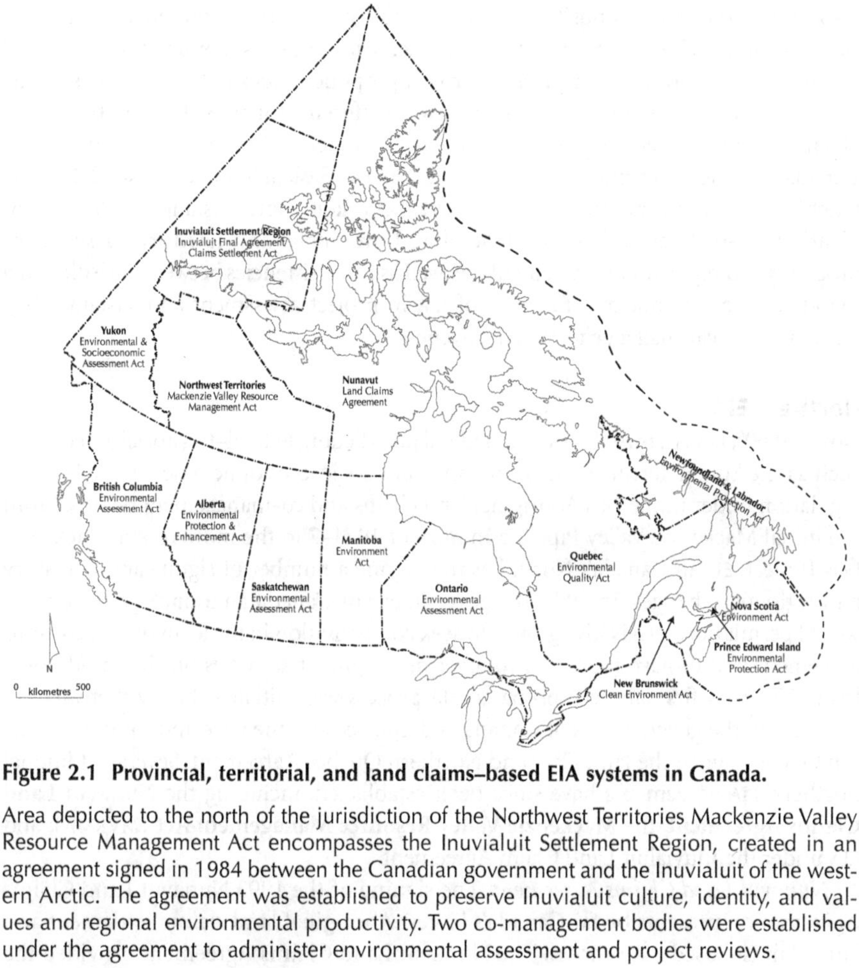

Figure 2.1 Provincial, territorial, and land claims–based EIA systems in Canada.

Area depicted to the north of the jurisdiction of the Northwest Territories Mackenzie Valley Resource Management Act encompasses the Inuvialuit Settlement Region, created in an agreement signed in 1984 between the Canadian government and the Inuvialuit of the western Arctic. The agreement was established to preserve Inuvialuit culture, identity, and values and regional environmental productivity. Two co-management bodies were established under the agreement to administer environmental assessment and project reviews.

Source: Map produced by J. Bronson, University of Saskatchewan.

Provincial EIA

Canada's first provincial EIA system, Ontario's *Environmental Assessment Act*, was enacted in 1975. All provinces and territories now have some form of EIA regulatory approval process, whether EIA requirements under environmental or resource management law, such as New Brunswick's *Clean Environment Act* and Newfoundland's *Environmental Protection Act*, or requirements under EIA-specific law, such as Saskatchewan's *Environmental Assessment Act*. In 1998, to facilitate co-ordination of EIA applications and responsibilities between the government of Canada and the provinces and territories, the Canadian Council of Ministers of the Environment signed an accord designed to improve cooperation on environmental assessment and protection. The purpose of the accord is to streamline the EIA process and eliminate duplication between federal and provincial or federal and territorial processes in cases where a proposed project could potentially trigger dual EIA systems. The intent of harmonization is not to transfer jurisdictional authority from the federal government to provincial or territorial governments but rather to ensure that federal and provincial or territorial approval processes are co-ordinated such that a project proponent does not need to present the same impact statement twice or in two different formats (Meredith 2004). Harmonization agreements on EIA have been signed between the government of Canada and the governments of Manitoba, Alberta, Saskatchewan, British Columbia, Quebec, Newfoundland and Labrador, and the Yukon. Notwithstanding EIA harmonization, co-ordination between federal and provincial and territorial assessment processes and regulatory requirements, as well as inconsistencies between jurisdictional mandates over the role and function of EIA in project assessment and sustainability, remains a critical challenge to EIA effectiveness.

Northern EIA

North of 60°, EIA is a mixed system of federal jurisdiction, federal–territorial agreements such as the Yukon Territory Environmental and Socio-economic Assessment Act, and regulation under numerous Aboriginal land claims and co-management boards. From the initial Mackenzie Valley Pipeline Inquiry of 1974–7 to the more recent Mackenzie Gas Project, EIA in Canada's North has undergone a number of significant regulatory and legislative changes. In 1973, the government of Canada announced a policy that would permit northern Native groups to seek compensation in the form of a land claims agreement and to have more control over development activities on their traditional lands. The very first land claims–based EIA process was initiated shortly thereafter, in 1975, when the governments of Canada and Quebec and the Cree and Inuit of northern Quebec signed the James Bay and Northern Quebec Agreement. Several additional northern EIA agreements have since been established, including the **Nunavut Land Claims Agreement**, the **Mackenzie Valley Resource Management Act** (MVRMA), and EIA under the Inuvialuit Land Claim Agreement.

Nunavut Land Claims Agreement. The signing of the 1993 Nunavut Land Claims Agreement represents by far Canada's largest Aboriginal land claims settlement and land claims–based EIA process. The Nunavut Land Claims Agreement provided the Inuit with self-government title to approximately 350,000 square kilometres of land in what had been the eastern half of the Northwest Territories. The negotiated settlement area included mineral rights for approximately 35,000 square kilometres of this land. The agreement also included an undertaking for the implementation of a new territory, Nunavut, which was formally established in 1999 under the agreement (Figure 2.1). Included under Article 12.2.2 of the agreement is the establishment of an environmental assessment review board, the **Nunavut Impact Review Board** (NIRB), which is the primary authority responsible for EIA activities in the land claims area. The mandate of the NIRB includes:

1. screening proposals to determine whether an environmental review is required;
2. gauging and determining the extent of the regional impacts of proposed projects;
3. reviewing ecosystemic and socio-economic impacts of proposed projects;
4. determining whether proposed projects should proceed and under what conditions;
5. monitoring development projects (see www.ainc-inac.gc.ca).

The overall purpose of the NIRB is to protect and promote the future well-being of residents and communities of the Nunavut settlement area, including its ecosystem integrity.

Mackenzie Valley Resource Management Act. For the purposes of the MVRMA, the Mackenzie Valley is defined as including all of the Northwest Territories, with the exception of Wood Buffalo National Park and the Inuvialuit Settlement Region (Figure 2.1). Proclaimed in 1998, the MVRMA was implemented by the federal government with the intent of giving northerners increased decision-making authority over natural resource and environmental development matters. A co-management board was also established for the Sahtu and Gwich'in settlement areas of the Northwest Territories to delegate responsibilities for land-use planning and for issuing land-use permits and water-use licences. As part of this establishment, a Valley-wide public board, the **Mackenzie Valley Environmental Impact Review Board** (MVEIRB), was created to undertake EIAs and panel reviews within the jurisdiction of the MVRMA. The Canadian Environmental Assessment Act no longer applies in the Mackenzie Valley except for very specific situations involving issues that are transboundary in nature or for which it is deemed necessary by the board and the federal minister of environment. In such cases, a joint MVEIRB-CEAA review panel is established. The MVEIRB is responsible for:

1. conducting environmental impact assessments;
2. conducting environmental reviews;
3. maintaining a public registry of environmental assessments;
4. making recommendations to the minister of the Department of Indian and Northern Affairs regarding the approval or rejection of development projects within the region.

Inuvialuit Land Claim Agreement. In 1984, a comprehensive land claim, the Inuvialuit Final Agreement (IFA), was signed between Canada and the Inuit of the western Arctic (Figure 2.1). The IFA encompasses approximately 1,000,000 square kilometres, known as the Inuvialuit Settlement Region, and includes six communities. One of the outcomes of the IFA was the establishment of two co-management agencies, the Environmental Impact Screening Committee and the Environmental Impact Review Board, to deal with environmental assessment screenings and reviews in the Inuvialuit Settlement Region. During the agencies' first 10 years, more than 150 development proposals were screened and two public reviews undertaken for offshore oil and gas exploration programs. In 2000, Canada and the Inuvialuit Environmental Impact Review Board signed an environmental assessment agreement for the Inuvialuit Settlement Region. The agreement concerns how the EIA process under the Environmental Impact Review Board will be substituted for a panel review under the Canadian Environmental Assessment Act and details the process and the steps involved should the Environmental Impact Review Board request such a substitution.

LEGAL DEVELOPMENT OF EIA

> Environmental assessment over the last 30 years has moved towards being earlier in planning, more open and participative, more comprehensive, more mandatory, more closely monitored, more widely applied, more integrative, more ambitious, and more humble (Gibson 2002, 152).

The current EIA regulatory system in Canada is the product of nearly 40 years of development, legislative reform, and political controversy. EIA at the federal level is currently required by means of EIA-specific law, and it is the responsibility of the proponent (the company or individual proposing the development) to undertake an EIA and prepare an EIS as per federal, territorial, and provincial regulations and requirements. When EIA was first formally introduced in Canada in 1973, it was never the intent of the federal government that EIA would someday have a legal basis.

EARP Origins

In Canada, EIA originated with the establishment of a task force in 1970 to examine a policy and procedure for assessing the impacts of proposed developments. In 1973, the Cabinet Committee on Science, Culture and Information agreed on the need for a formal process to assess the potential impacts associated with project development. At that point, the first Canadian EIA process, the federal **Environmental Assessment Review Process** (EARP) Guidelines Order, was born and the **Federal Environmental Assessment Review Office (FEARO)** was created to administer its implementation. Unlike NEPA in the US, however, the Canadian EARP had no legislative basis and was therefore not legally enforceable; rather, project impact reviews were initially intended to be cooperative and voluntary, not legally binding. The result was that impact assessments under EARP were carried out inconsistently and in some cases not carried out at all. Some of the earliest projects reviewed under EARP included the Point Lepreau nuclear power station (1975), eastern Arctic offshore drilling (1978), and the Arctic Pilot Project (1980). Perhaps the most significant project reviews under EARP, reviews that would change the course of EIA in Canada, were the Oldman River Dam project in Alberta (Box 2.1) and the Rafferty-Alameda Dam in Saskatchewan—both initiated without formal project assessment.

Canadian Environmental Assessment Act

Introduced in Parliament in 1992 by the Conservative government as Bill C-78, the **Canadian Environmental Assessment Act** was created to replace EARP. While the new Act was intended to make EIA more rigorous and systematic, it also limited the reach of EIA to include only project-level decisions and not broader planning or policy issues. The Act set out responsibilities and procedures for the environmental assessment of projects involving federal authorities and established a process to ensure that impact assessment was applied early in the planning stages of proposed project developments. When the Act applies to a proposed project, the relevant federal government department is designated as the 'responsible authority' and must ensure that EIA unfolds according to the principles of the Act.

In 1994, the **Canadian Environmental Assessment Agency** was created to oversee the Act and replaced FEARO. The Act was proclaimed and came into force in 1995. Its purposes are:

1. to ensure that projects are considered in a careful and precautionary manner before federal authorities take action in connection with them so that such projects do not cause significant adverse environmental effects;

2. to encourage responsible authorities to take actions that promote sustainable development and thereby achieve or maintain a healthy environment and a healthy economy.

The Act itself has several main objectives, namely:

- to ensure that the environmental effects of projects are reviewed before federal authorities take action so that potentially significant adverse impacts can be ameliorated;
- to encourage federal authorities to take actions that promote sustainable development;
- to promote cooperation and co-ordinated action between federal and provincial governments on environmental assessments;
- to promote communication and cooperation between federal authorities and Aboriginal peoples;
- to ensure that development in Canada or on federal lands does not cause significant adverse environmental effects in areas surrounding the project;
- to ensure that there is an opportunity for public participation in the environmental assessment process.

When the Act was initially created, section 72 required that a review of the Act be undertaken every five years. In 1999, Bill C-13, An Act to Amend the Canadian Environmental Assessment Act, was introduced, and public consultations on EIA reform were organized by the Standing Committee on the Environment and Sustainable Development. Following consultation, the standing committee reintroduced the bill to Parliament as Bill C-9, which received royal assent on 11 June 2003, and the revised Canadian Environmental Assessment Act became law on 30 October 2003 (see www.ceaa.gc.ca). Among the commitments under the new Act are:

1. Promotion of high-quality assessments. The revisions will potentially contribute to better decision-making in support of sustainable development by improving compliance with the Act, strengthening the role of follow-up in EIA, providing the minister of environment with the option of requesting follow-up studies or additional information before making a project decision, and improving the consideration of cumulative effects.
2. Assessment of appropriate projects. The revised Act is intended to focus the EIA process on projects likely to have significant adverse environmental effects while reducing the need for and time commitment to assessing smaller, routine projects. This involves making use of a more streamlined assessment process for routine projects and exempting certain smaller projects from EIA requirements provided that specified environmental conditions are met.
3. Application of the Act fairly and consistently. EIA now extends to reserve lands, federal lands, Crown corporations, and the assessment of transboundary effects that may occur on park reserve lands and areas of land claims.

Box 2.1 Reforming EIA: Oldman River Dam Project, Alberta

Throughout the 1960s and into the 1970s, agriculture was of growing importance in Alberta. Periodic droughts, characteristic of the region's semi-arid continental climate, meant considerable uncertainties in crop production and sustainability of local water supplies. Combined, agricultural growth and climatic unpredictability reinforced earlier initiatives for the development of large-scale irrigation systems. In 1976, plans were announced to construct an earth- and rock-filled dam at the Three Rivers site in Alberta (Oldman, Castle, and Crowsnest rivers). The proposed dam was to be approximately 75 metres in height and more than 3,000 metres long, with a reservoir capable of storing nearly 500 million cubic metres of water (Oldman River Dam Project 1992). Several sites for the dam were considered, but the Three Rivers site was favoured from the outset, notwithstanding studies that identified potentially significant adverse effects on fish and wildlife as well as impacts on ranchers whose properties would be partially divided and flooded.

The controversy over the proposed project occurred over two periods. First, between 1976 and 1980, landowners whose lands would be flooded opposed the dam, and the Peigan Indian Reserve argued that the project was intruding on lands within the Blackfoot Nation's territory. This first controversy in part ended in 1980 when the final project site was selected. The second, and more significant, controversy emerged in 1987 when an environmental group, Friends of the Oldman River, challenged the project through the provincial courts. The environmental group was initially unsuccessful, and based on assessments under Alberta's environmental assessment process and with the approval of the federal government under the Navigable Waters Protection Act, project construction commenced in 1988. No federal assessment under the EARP Guidelines Order was initiated, notwithstanding the involvement of a federal authority. The Oldman Dam project was nearly 40 per cent complete when Friends of the Oldman River once again challenged the project in 1989 but this time in the federal court. The environmental group lost the court case, but in an interesting turn of events, the court of appeals in 1992 reversed the initial decision and quashed approval for the Oldman Dam project under the Navigable Waters Protection Act. The court of appeals ordered that the EARP did in fact have the force of law, and the federal government was compelled to comply with the Guidelines Order. While the decision was too late to reverse any damage that had already been created by the dam project, it did pave the way for a legal requirement for EIA in Canada, which would ensure that EIA would be implemented before development took place.

4. Improvement of co-ordination among participants. To minimize duplication, the position of federal environmental assessment co-ordinator has been established to facilitate EIAs that involve more than one jurisdictional authority, such as federal–provincial EIA agreements.
5. Increased certainty in the process. In addition to clarification of key terms and procedures under the Act, the process will make greater use of mediation and dispute resolution to improve EIA efficiency. The Canadian Environmental Assessment Agency is also committed to playing a greater role in building consensus and co-ordinating the dissemination of information through early involvement in the assessment process.

6. Improvement of public participation. The nature and role of public participation is to be enhanced through improved access to information, including the federal registry of impact statements and reports, expanded opportunities for public input, and better incorporation of Aboriginal perspectives and values into the decision-making process.

At the time this book was written, the Act was again under federal review. Currently, the Act requires that an impact assessment be carried out when a federal authority is involved in a project. This involves cases when a federal authority—i.e., a ministry or agency of the federal government—proposes a project, provides financial assistance for carrying out a project, transfers control or administration of federal land to enable a project to be carried out, or provides certain licences or permits enabling a project to be carried out. This, combined with the current listed projects requiring federal EIA in Canada (see Chapter 5), results in a large number of federal impact assessments carried out on an annual basis—many of which are for fairly routine undertakings (see Table 2.1). While no formal decisions have yet been made as to the scope of EIA likely to emerge from the next iteration of the Act, there is considerable federal interest in and discussion about a significant streamlining of EIA—specifically, a reduction in the number and types of projects that would be subject to EIA at the federal level, with perhaps greater responsibility for project assessment devolving to the provinces and territories. Ideally, this streamlining of EIA at the project level would be balanced by increased support for broader regional scale and strategically based environmental assessment systems and practices (see Chapters 13 and 14).

Towards 'Higher-Order' Assessment

Perhaps the most significant advancement in Canadian EIA in recent years is the development of environmental assessment above the project level—at the level of policies, plans, and programs. In Canada, the environmental assessment of policies was included in the former EARP Guidelines Order, which defined 'proposal' as any initiative for which the federal government had decision-making responsibility. With the introduction of the 1992 Canadian Environmental Assessment Act, however, environmental assessment was restricted by law to 'projects' or 'physical undertakings', and policy, plan, and program assessment was required only by means of federal cabinet directive.

In 1990, Canada announced a reform package for environmental assessment that included new legislation and a new assessment process for decisions and actions above the project level. The commitment to strategic environmental assessment was strengthened in 1999 with release of the cabinet directive on the Environmental Assessment of Policy, Plan and Program Proposals. Under the directive, an environmental assessment of a policy, plan, or program is required when a proposal is submitted to a minister or to cabinet for approval and implementation of that proposal may result in either positive or negative environmental effects. The overall purpose of this strategic approach to environmental assessment is to allow for more informed decisions in support of sustainability initiatives. The directive was updated in 2004 (see Chapter 13).

EIA DIRECTIONS

Initiated as a policy with little grip, EIA has since evolved to become a rigorous regulatory and legal framework. How this evolution has unfolded over the past 35 years, however, might be described either as the dislocation of an established order in order to improve regulatory requirements, as the natural transition from one system to another, or as a series of unnatural breaks categorized by forces of social and legal change. However one chooses to categorize the history of EIA in Canada, Gibson (2002, 156) captures it best in suggesting that 'Canadian environmental assessment policies and laws have evolved slowly . . . this evolution has been hesitant and uneven, though overall it has been positive.'

KEY TERMS

Canadian Environmental Assessment Act
Canadian Environmental Assessment Agency
Environmental Assessment Review Process
Federal Environmental Assessment Review Office (FEARO)
Mackenzie Valley Environmental Impact Review Board
Mackenzie Valley Resource Management Act
Nunavut Impact Review Board
Nunavut Land Claims Agreement

STUDY QUESTIONS AND EXERCISES

1. What are some of the critical events that have shaped the course of Canadian EIA at the federal, provincial, and territorial levels?
2. When was EIA enacted in your province (or territory)? What is the name of the legislation that requires EIA? Describe some of the main features of your province or territory's EIA legislation, such as its purpose, how it defines 'environment', and when an EIA is required. Compare this to the EIA legislation of another Canadian province or territory.
3. It has been suggested that all of Canada should be governed by a single EIA regulatory framework and assessment process. Do you agree? What might be the advantages and challenges to removing EIA responsibility and authority from provinces and territories and transferring it to a single, composite federal EIA system?

REFERENCES

Bitter, B. 2008. Personal communication (8 October 2008). Environmental Assessment Branch, Ministry of Environment, Government of Saskatchewan.

Gibson, R.B. 2002. 'From Wreck Cove to Voisey's Bay: The evolution of federal environmental assessment in Canada'. *Impact Assessment and Project Appraisal* 20 (3): 151–9.

Hanna, K., Ed. 2009. *Environmental Impact Assessment: Practice and Participation*. 2nd edn. Don Mills, ON: Oxford University Press.

Meredith, T. 2004. 'Assessing environmental impacts in Canada'. In B. Mitchell, Ed., *Resource and Environmental Management in Canada*, 3rd edn, 467–96. Toronto: Oxford University Press.

Oldman River Dam Project. 1992. *Report of the Environmental Assessment Panel*. Report no. 42. Hull, QC: FEARO.

PART II

Foundations of Environmental Impact Assessment

CHAPTER 3

Nature and Classification of Environmental Effects

ENVIRONMENTAL CHANGE AND EFFECTS

The terms 'change' and 'effect' are often used interchangeably. In principle, however, an environmental change is a temporal measure, and not all change is project-induced; an environmental effect is the difference between change conditions. A distinction can thus be made between environmental change and environmental effects in impact assessment. **Environmental change** is the difference in the condition of a particular environmental or socio-economic parameter, usually measurable, over a specified period of time. Environmental change is typically defined in terms of a process, such as soil erosion, that is set in motion by particular project actions, other actions, or natural processes (Figure 3.1). Actions such as road construction or dam construction, for example, contribute to environmental change. An **environmental effect** is the change difference: the difference in the condition of an environmental parameter under project-induced change versus what that condition might be in the absence of project-induced change.

The term environmental effect is also often used interchangeably with **environmental impact**; however, some argue that impacts are estimates or judgments of the value that society places on certain environmental effects or concerns about the dif-

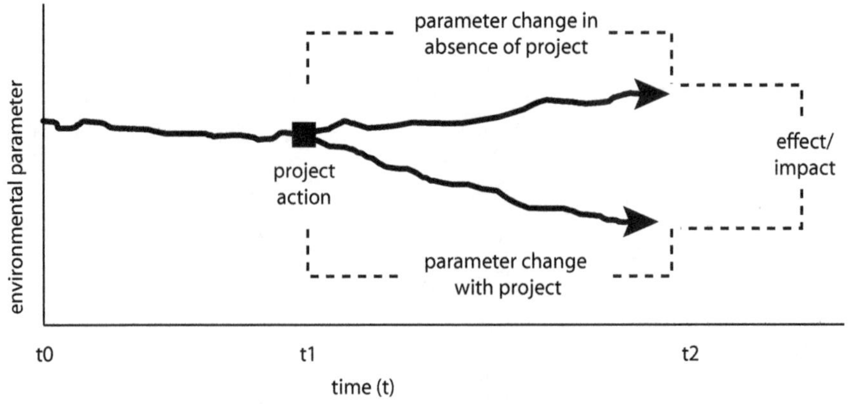

Figure 3.1 Conceptualization of an environmental effect.

ference between the quality of the environment with and without the proposed action. While some consider this distinction little more than a matter of semantics, the distinction between change and effects or impacts is particularly important from an environmental management perspective. For example, managing environmental 'impacts' may be no more than a damage control exercise if one does not successfully identify environmental effects and, perhaps even more important, the project-induced environmental change that is generating the effect or impact from the onset.

Order of Environmental Effects

Effects can be either **direct effects** (first order) or **secondary effects** (resulting from the direct effects) (Figure 3.2). For example, the flooding of land during construction of a hydroelectric project has a direct effect on terrestrial habitat. Indirectly, human health may be affected as a result of exposure to increased mercury levels in fish caused by mercury release and the flooding of soils and decay of organic matter.

The relationship between first-order effects (or impacts) and second-order effects (or impacts), however, is not always straightforward. For example, runoff from an agricultural operation may cause enrichment of a local freshwater body by depositing excess nutrients. The result may be excess plant growth near the shores of the water body, which may create an aesthetically pleasing waterscape. However, the subsequent eutrophication—excess plant decomposition absorbing dissolved oxygen at high rates—generates a negative secondary effect in making water unsuitable for aquatic life. Preston and Bedford (1988) suggest that an 'effect' is a scientific assessment of facts and an 'impact' denotes a value judgment or the relative importance of the effect (Box 3.1).

CLASSIFICATION OF EFFECTS AND IMPACTS

In addition to spatial and temporal scales, as introduced in Chapter 1, effects and impacts can be classified in a number of ways. In the Mackenzie Gas Project in Canada's Northwest Territories, for example, environmental effects and impacts are classified based on their direction, magnitude, spatial extent, and duration (Box 3.2). In short, there is no single best set of criteria that can be used to classify all

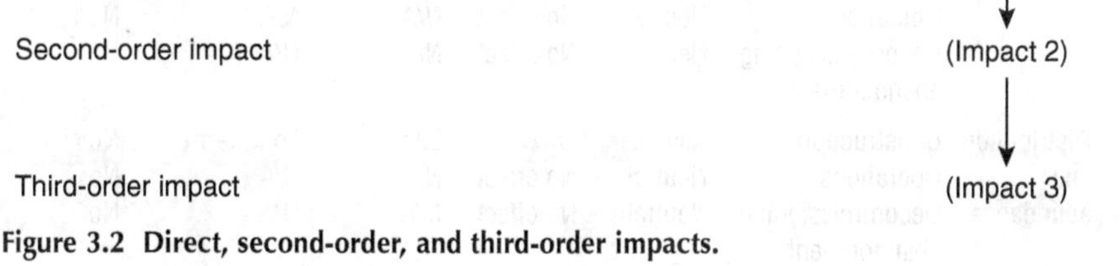

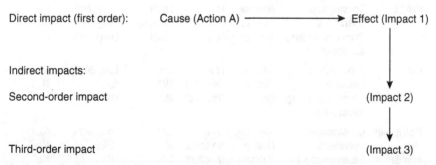

Figure 3.2 Direct, second-order, and third-order impacts.

Source: Based on Harrop and Nixon 1999.

Box 3.1 Actions, Causal Factors, Effects, and Impacts

Action	Hydroelectric dam construction on a river system. Assume a simple development project in which a river is modified and dammed for hydroelectric power generation.
Causal Factor	Dredging. The project involves dredging and widening the river channel upstream of the dam site prior to its construction.
Condition Change	Increased sediment flow into the river downstream of the dredging site creates changes in deposition of sediments and increased turbidity.
Effects and Impacts	Direct: Decreased growth rates in fish. Indirect: Decline in recreational fishery. Several different impacts may occur as a result of the initial change in condition. For example, deposition of sediments and increased turbidity (project effects) may result in decreased growth rates in river fish (first-order, direct impact), leading to a decline in the region's recreational river fishery (second-order, indirect impact).

environmental effects and impacts in every project situation, but there are several common classifications of effects and impacts that should be examined for each environmental component affected by a project (Table 3.1). The following is by no

Box 3.2 Effects of Natural Gas–Gathering Pipelines and Facilities on Fish and Fish Habitat for the Mackenzie Gas Project, Northwest Territories

		Effect attribute				
Key indicators	*Phase when impact occurs*	*Direction*	*Magnitude*	*Geographic extent*	*Duration*	*Significant*
Habitat	Construction	Adverse	Low	Local	Short term	No
	Operations	Adverse	Low	Regional	Long term	No
	Decommissioning/ abandonment	Adverse	Low	Local	Long term	No
Health	Construction	Adverse	Low	Local	Long term	No
	Operations	Neutral	No effect	N/A	N/A	No
	Decommissioning/ abandonment	Neutral	No effect	N/A	N/A	No
Distribution and abundance	Construction	Adverse	Low	Local	Long term	No
	Operations	Neutral	No effect	N/A	N/A	No
	Decommissioning/ abandonment	Neutral	No effect	N/A	N/A	No

continued

Effect attributes:

Direction

Adverse: Impact will cause an adverse change in a measurable parameter relative to baseline conditions or trends.
Neutral: Impact will cause no change in a measurable parameter relative to baseline conditions or trends.
Positive: Impact will cause a positive change in a measurable parameter relative to baseline conditions or trends.

Magnitude

No effect: No change in the valued component.
Low: An individual or group within a population found in a localized area, such as the local or regional study area, might be affected.
Moderate: Part of a regional population within the local or regional study area might be affected, changing the abundance or distribution of the valued component and affecting opportunities for hunting, trapping, or viewing wildlife as currently practised.
High: An entire population within the local or regional study area might be affected, changing the abundance or distribution to such an extent that the population would not likely return to its previous level, resulting in reduced population viability and unsustainable harvest compared with current practice.

Geographic extent

Local: Terrestrial—the effect on the valued component is measurable within the local study area; marine—the effect will be limited to within about 10 kilometres of the proposed activity.
Regional: Terrestrial—the effect on the valued component is measurable within the regional study area; marine—the effect will extend beyond 10 kilometres of the proposed activity to the Canadian Beaufort Sea region.
Beyond regional: Terrestrial—the effect on the valued component is measurable beyond the regional study area.

Duration

Short term: Effect is limited to less than one year.
Medium term: Effect lasts for more than one year but less than four years.
Long term: Effect lasts longer than four years, but the valued component will recover not more than 30 years after project decommissioning.
Far future: Effect extends more than 30 years after decommissioning.

Source: Mackenzie Gas Project Environmental Impact Statement, 2004, v. 5, section 7, pp. 7-183 to 10-27.

means a comprehensive list; rather, it serves to illustrate the variety of effects and impacts that can be identified and assessed in an EIA.

Nature of Environmental Effects and Impacts

The nature of the actual effect or environmental impact can be divided into incremental, additive, synergistic, and antagonistic categories.

Incremental effects or impacts are marginal changes in environmental conditions that are directly attributable to the action being assessed. For example, a hydroelectric

Table 3.1 Generic Classification System for Environmental Effects and Impacts

Nature of effect or impact	*Temporal characteristics*	*Magnitude*	*Direction of change in the affected environmental parameter*	*Spatial extent*	*Reversibility of change of the affected environmental parameter*	*Probability of occurrence*
a) Incremental	a) Duration	a) Size	a) Increasing	a) On-site	a) Reversible	a) Likelihood
b) Additive	b) Continuity	b) Degree	b) Decreasing	b) Off-site	b) Irreversible	b) Risk
c) Synergistic	c) Immediacy	c) Concentration	c) Positive	d) Regional		
d) Antagonistic	d) Frequency		d) Negative			
	e) Regularity					

dam may lead to a slight increase in heavy metal concentrations in the river each year. For incremental effects, it is important to consider the 'rate of change' in the affected environmental parameter and to regularly assess total change against determined thresholds, standards, or specified targets.

Additive effects or impacts are the consequence of separate or related actions that may be minor individually but together can create a significant overall impact (Figure 3.3). For example, chemical 'A' may result in 10 per cent mortality in fish over a 30-day exposure period. Chemical 'B' may result in the same mortality over the same time period. If chemical 'A' and chemical 'B' were released together, the additive mortality in fish would be 20 per cent. Additive effects are not only the result of progressive increases in environmental parameters but also include, for example, the actions of several large-scale disturbances, such as multiple logging operations, within a single ecosystem. Thus, while perhaps individually manageable, together such operations present significant adverse additive effects on wildlife because of habitat fragmentation. Additive effects may therefore result from two types of actions:

- Where two or more of the same type of actions are affecting the same environmental component. For example, multiple coal-bed methane developments may generate cumulative effects on the landscape as a result of saline water discharge.
- Where a single development action produces multiple condition changes or effects that result in effects on the same environmental component. For example, the development of a mine site may create habitat loss, changes in surface water drainage and quality, and increased noise and emissions from vehicles and operations. In combination, they can have a cumulative negative effect on wildlife.

Synergistic effects or impacts are the result of interactions between effects and occur when the total effect is greater than the sum of the individual effects (Figure 3.4). For example, a single industry located adjacent to a river system may alter water temperatures, change dissolved oxygen levels, and introduce heavy metals. Individually, each effect may be tolerable for fish, but the toxicity of certain heavy metals is multiplied in high water temperatures and low dissolved oxygen content. The effect on fish as a result of the interaction of these effects is thus greater than the sum of the individual project-induced changes. Synergistic and additive effects are discussed in greater detail in Chapter 12 on cumulative environmental effects.

Antagonistic effects or impacts occur in certain situations in which one adverse effect or impact may partially cancel out, offset, or interrupt another adverse effect or impact. For example, when chemical 'A' with a 10 per cent mortality rate in fish is

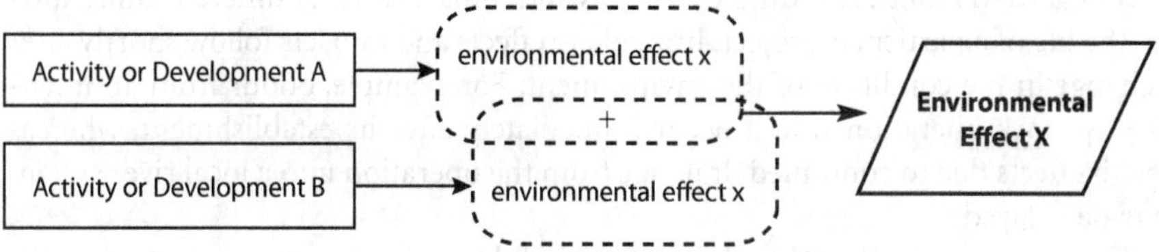

Figure 3.3 Additive environmental effects.

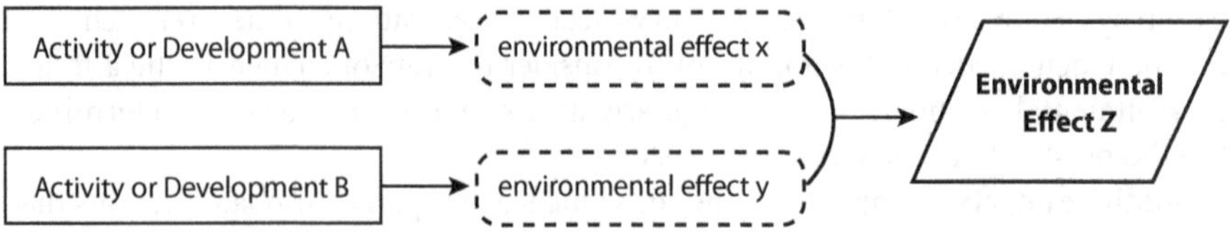

Figure 3.4 Synergistic environmental effects.

combined with chemical 'B', which has the same mortality rate, the antagonistic result is less than a 20 per cent mortality rate. These effects are usually less common in EIA, since additional stressors to the environment more often create further disturbance and degradation. One example, however, is the reduced eutrophication of a water body receiving effluents containing both chlorine and phosphates. While each substance is individually harmful for aquatic life, together and in moderate amounts they may be beneficial for managing eutrophication.

In terms of the biophysical environment, antagonistic effects can be further defined according to functional, chemical, dispositional, and receptor antagonism (Kabata-Pendias 2000). **Functional antagonism** refers to one effect counterbalancing the physiological response of a second effect on the same receptor. **Chemical antagonism**, on the other hand, refers to a reaction between the two effects, such as in a chemical reaction, where the severity of the combined effect is reduced. For example, the total effects of arsenic and mercury are reduced when combined with dimercaprol, which itself is a toxic compound. In the case of **dispositional antagonism**, one effect influences the uptake or transport of the other, such as ethanol enhancing mercury elimination in mammals. **Receptor antagonism** refers to one effect blocking the other, such as a toxicant binding to a receptor and blocking the effects of a second toxicant.

Temporal Characteristics of Effects and Impacts

Since effects and impacts often occur over time, their characteristics may vary. Thus, for any effect or impact, we can characterize its temporal nature according to the following factors.

Duration is the length of time that the effect or impact occurs—for example, short-term effects such as the noise associated with a bridge construction site or the long-term effects associated with riverbank erosion and downstream loss of arable land.

Continuity relates to whether or not there are **continuous effects** or impacts, such as energy fields associated with transmission lines. Other effects may be discontinuous and last only for short intervals, such as the noise from blasting at a construction site.

In regard to *immediacy*, different effects and impacts arise at different times during the life of a particular project. Immediate effects and impacts follow shortly after a change in the condition of the environment. For example, odour from an intensive livestock operation would occur immediately after its establishment, whereas health effects due to continued drainage from the operation into a local river system may be delayed.

The *frequency* of a change in the condition of the environment may create very different impacts. For example, local neighbourhood residents may be willing to endure short-term noise from local street maintenance activities, but they may not be willing to tolerate the construction and operation of a new airport near their homes.

Effects or impacts that occur with *regularity* and greater predictability may often be dealt with more easily than those that happen irregularly or that come as a surprise. For example, the 'startle effect' of low-level military flight training in Labrador on local Aboriginal populations and caribou resources is of particular concern, not because of the noise level itself but because of the surprise effect of low-flying military jets (Box 3.3).

Magnitude, Direction

The **magnitude** of an effect refers to its size or degree—for example, the specific concentration of an emitted pollutant. That said, while many impacts can be measured in quantitative terms such as concentration or volume, others, such as the aesthetic impact on a natural area, are subjective matters and not easily indexed or reduced to absolute quantitative terms. It is important to note that the magnitude of an effect

Box 3.3 Regularity and the Startle Effect: Low-Level Military Flight Training in Labrador

Canada and its NATO allies have been engaged in low-level military flight training in Labrador since the early 1980s. Approximately 6,000 to 7,000 flights are conducted annually, all less than 1,000 feet above the surface. Ninety per cent of these flights are below 500 feet, and approximately 15 per cent are as low as 100 feet. Churchill Falls is the only community within the vicinity of flight training activities. However, about a dozen small Innu settlements are located within the area, and these Aboriginal people rely heavily on hunting and gathering activities, particularly caribou harvesting, within the training area.

In 1990, after emerging concerns over the potential impacts associated with low-level flying activities, the Canadian Department of National Defence (DND) initiated several monitoring programs, including monitoring the potential impacts of noise on wildlife and humans. An EIA was completed in 1994 and subsequently approved in 1995 for continued flight training. A number of additional wildlife monitoring programs were implemented as a result of the EIS, particularly for the Georges River caribou herd—an important subsistence resource base for local Aboriginal communities, with approximately 38,000 caribou harvested annually. Of particular concern was noise—not noise levels per se but the 'startle effect' or unpredictability of low-flying military jets.

The rate at which noise increases is referred to as the onset rate. Onset rates greater than 15 decibels (dBA) per second are considered 'startling' according to DND reports. Onset rates of low-level military flights can often exceed 150 dBA per second. While the decibel rating of the noise level itself is considered tolerable, there is some concern that the unpredictability of the impact of overhead jets may affect the physiological conditions of the caribou, including reproduction performance and patterns of habitat use. While DND received considerable input from experts and local residents concerning the impact of noise on caribou, there was no clear resolution of the actual significance of the startle effect.

Source: Based on CEAA 1995.

is not necessarily related to the significance or importance of the impact. For example, removal of 5 per cent of the forest cover in an area that supports a rare or endangered species may be considered much more significant than removal of 25 per cent of forest cover in an area that supports stable native species populations.

The *direction* of change of an effect is closely associated with its magnitude. Direction of change refers to whether the affected environmental parameter is increasing or decreasing in magnitude relative to its current or previous state or **baseline condition**. For example, the construction and operation of a large-scale industry may lead to an increase in the local population if it generates employment opportunities and triggers in-migration; however, it may lead to a decrease in local population if smaller industries are displaced or if quality of life in the community decreases as result of environmental change. In the first instance, the impact may be considered positive, whereas in the second, an **adverse effect** is the more likely result. The situation is not always this straightforward. In some cases, effects are multi-directional. For example, some individuals in the affected community may see an increase in local population as a positive impact and an improvement in their quality of life, whereas others may see population increase due to industrial growth as having an adverse impact on their quality of life. When characterizing environmental change, it is important to give at least some indication of direction, even if it cannot be quantified, in order to better understand the magnitude of the effect and impact.

Spatial Extent

The *spatial extent* of an environmental effect or impact will vary depending on the specific parameters involved and the project's environmental setting. For example, soil erosion may be highly localized and agricultural drought may be regional, while economic impacts associated with continuous drought may be national or international in extent. In the case of the Jack Pine mine project, a proposal by Shell Canada to mine bitumen deposits in the Athabasca oil sands of Alberta (see Chapter 4, Box 4.7), the spatial extent of the assessment was defined by 'local study areas' and 'regional study areas'. Local study areas were defined as those directly affected by project development. Regional study areas were identified as those seen from a larger geographic and ecological perspective that would experience direct and indirect impacts.

Sometimes spatial extent is classified according to **on-site impacts** and **off-site impacts**, a recognition that the impacts of any particular development or action may extend well beyond the specific project site. On-site actions may trigger environmental change, which in turn affect off-site environmental components. For example, chemical discharges at an industrial complex generate on-site environmental impacts due to soil contamination, which in turn may create off-site impacts through groundwater contamination. Such off-site impacts may go beyond biophysical impacts alone. **Fly-in fly-out** employment arrangements, often associated with remote mining projects, have impacts that extend well beyond the mine project site to the communities the workers come from and affect personal matters such as family life and relationships. Determining the spatial extent of EIA is a key element of the scoping process, discussed in greater detail in Chapter 6, and is critical to ensuring that all necessary impacts are considered within the scope of the assessment.

Reversibility and Irreversibility of Effects and Impacts

An additional characteristic to consider when examining environmental effects and impacts is the degree of reversibility or irreversibility. This is particularly important for determining the nature and effectiveness of impact management strategies.

An effect or impact is considered *reversible* when it is possible to approximate the pre-disturbed condition. For example, in the case of a sanitary landfill operation, the surface can be remediated to resemble its pre-project condition, or traffic congestion may return to its original level when a bridge reconstruction project ends. Reversing an environmental effect, however, does not necessarily mean restoring a biophysical or socio-economic environment to its *exact initial* condition, since that is neither always possible nor always desirable. A mine site, for example, may be remediated and reforested based on current forest stand composition in the project area because it may neither be desirable nor feasible to return the site to its pre-project forest stand composition. The environment is a moving target; it changes and adapts irrespective of project actions. Restoring disturbed environments to their pre-disturbed condition is at best an approximation of what the affected environment might look like had the development action not taken place.

Some effects and impacts are completely *irreversible* and are thus of considerable importance when determining significance. For example, the extinction of a rare plant species during site-clearing is irreversible. Other types of impact may be reversible from a technical standpoint, but it may be impractical or economically unfeasible to attempt to do so. For example, it is common practice in large-scale mining initiatives to drain a local pond or lake for tailings disposal. While it is *possible* to restore the pond or lake following mine decommissioning, restoration is often neither practical nor economically feasible. Consequently, these effects are often considered irreversible from any practical impact management perspective. Instead, efforts are made elsewhere to compensate for the loss of aquatic habitat, either through the creation of new habitat or the restoration of damaged habitat. Although these effects are considered irreversible, that does not mean that there is no need for managing such impacts or for site cleanup. The failure to perform site restoration and cleanup following project operations has been particularly problematic in Canada's mining sector (Box 3.4).

Likelihood of Effects and Impacts

Whether an environmental effect or impact is considered important during the course of an EIA rests considerably on the likelihood of it occurring. If, for example, a particular impact is almost certain to occur, then it is more likely to be considered in the decision-making process than one that is determined to be highly improbable. The likelihood or probability of an effect or impact occurring is one component used to measure **risk**—an uncertain situation involving the possibility of an undesired outcome. Risk combines the likelihood of an adverse event occurring with an analysis of the severity of the consequences associated with that event. In other words, risk can be characterized as a function of damage potential, exposure to dangerous substances such as toxic chemicals, opportunity for exposure, and the characteristics of the population at risk. For example, the elderly and young children

Box 3.4 Reclaiming Abandoned Mine Sites in Northern Canada

The Lorado uranium mine/mill site, located approximately eight kilometres south of Uranium City in northern Saskatchewan, ceased operation in 1960. The mine site was owned by what is now EnCana West Ltd, a Calgary-based company and formerly AEC West Ltd, renamed after the 2002 merger of PanCanadian Energy Corporation and Alberta Energy Company Ltd. The mine pre-dated any formal requirements for EIA or post-closure site reclamation, and as a result, left behind are exposed tailings—waste rock left over after the ore is extracted—containing metals, radioactive elements, and generate acid. The Lorado mine is just one of 65 abandoned mine sites in northern Saskatchewan for which no corporation or government agency is clearly responsible for site restoration and cleanup.

Saskatchewan's Lorado mine site is not a unique case; in fact, thousands of abandoned mine sites in northern Canada have not been reclaimed and continue to pose serious threats to the environment and to human health and safety. In the past, little could be done to ensure that mine sites were reclaimed or to prevent bankrupt companies from 'walking away' from their operations. As a result, the federal government currently spends millions of dollars each year to contain pollutants that are left behind at abandoned mine sites. In 2002, for example, the prevention of water contamination from abandoned mine sites cost Canadian taxpayers an estimated $25 million (Commissioner of the Environment and Sustainable Development 2002).

While the burden for cleanup and restoration of currently abandoned mine sites rests with the various federal and provincial governments, longer-term and viable solutions for environmental protection in Canada's mining industry now exist. Under the Canadian EIA system and most provincial systems, mining companies are responsible for preparing mine closure plans for approval for both new and active mines and, in addition, must provide post-closure financial assurance for environmental rehabilitation or stabilization. For mining companies operating in Canada's North, a financial security deposit is required at the time of project start-up and during operation. Should a mining company declare bankruptcy, the Department of Indian Affairs and Northern Development, the department responsible for land administration in northern Canada on behalf of the federal government, will use the security deposit to cover the eventual costs of repair, maintenance, cleanup, and closure of the mine site. If a mining company conducts the proper cleanup and site reclamation, the financial security deposit is returned.

may be much more vulnerable to exposure than other segments of the population. **Risk assessment**, then, simply refers to the process of accumulating information, identifying possible risks, risk outcomes, and the significance of those outcomes, and assessing the likelihood and timing of their occurrence. The acceptability of a specified level of risk depends on several factors, including the catastrophic potential associated with the risk event, scientific uncertainty, distribution of risk outcomes, and understanding of and familiarity with the risk. In terms of the latter, for example, an individual living on a flood plain may have perceptions of flood risk significantly different from those of individuals living at a distance from the same flood plain.

KEY TERMS

additive effects
adverse effects
antagonistic effects
baseline condition
chemical antagonism
continuous effects
direct effects
dispositional antagonism
environmental change
environmental effects
environmental impacts
fly-in fly-out
functional antagonism
incremental effects
magnitude
off-site impacts
on-site impacts
receptor antagonism
risk
risk assessment
secondary effects
synergistic effects

STUDY QUESTIONS AND EXERCISES

1. Consider a proposal for the construction of an industrial complex that requires forest clearing of the proposed site adjacent to a river system. Following the example presented in Box 3.1, identify potential causal factors, condition changes, and direct and indirect impacts of the proposed action.
2. Provide an example for each of incremental, additive, and synergistic impacts.
3. Suppose a proponent submits an application for the development of a large-scale open-pit gold mine operation in your region. In small groups, brainstorm the potential 'on-site' and 'off-site' impacts. Use the 'impact characteristics' discussed in this chapter to identify the impacts you believe are most important to consider among both on-site and off-site impacts. Are there any additional criteria or characteristics that should be considered? Compare the results among groups.

REFERENCES

CEAA (Canadian Environmental Assessment Agency). 1995. *Military Flying Activities in Quebec and Labrador*. Report of the Environmental Assessment Panel. Ottawa: Minister of Supply and Services Canada.

Commissioner of the Environment and Sustainable Development. 2002. *Report of the Commissioner of the Environment and Sustainable Development*. Chapter 3, 'Abandoned mines in the North'. Ottawa: Office of the Auditor General of Canada. www.oag-bvg.gc.ca/domino/reports.nsf/html/c2002menu_e.html.

Kabata-Pendias, A. 2000. *Trace Elements in Soils and Plants*. 3rd edn. Boca Raton, FL: CRC Press.

Preston, E., and B. Bedford. 1988. 'Evaluating cumulative effects on wetland functions: A conceptual overview and generic framework'. *Environmental Management* 12 (5): 565–83.

CHAPTER 4

Methods Supporting EIA Practice

METHODS AND TECHNIQUES

Literature on impact assessment often tends to use the terms 'methods' and 'techniques' in an imprecise way, treating them as synonymous. Such confusion in terminology often gives the impression that 'methods', such as assessment matrices or checklists, are used to predict environmental impacts. This is not so. **Methods** are concerned with the various aspects of assessment, such as the identification and description of likely impacts and the collection and classification of data; techniques provide those data (Barrow 1997; Canter 1996; Bisset 1988).

Techniques provide data that are then collated, arranged, analyzed, presented, and sometimes interpreted according to the organizational principles of the methods being used (Bisset 1988). A technique, such as a **Gaussian dispersion model**, provides data on some parameter, such as the anticipated dispersion of air pollutants from a specific industrial development; those data are then organized according to a particular method, such as an assessment matrix, in which the researcher evaluates and presents the data.

In any single EIA, a number of techniques and methods may be used, but techniques and methods differ for different types of assessment. Assessment techniques, which provide the data, are much more selective than assessment methods. Many of the techniques adopted for project- and program-level assessments, such as pollutant dispersion models or population forecasting models, are not appropriate at the policy level where the issues are by nature much broader. On the other hand, many surveying and forecasting techniques based on the use of expert judgment are equally applicable to the policy and project levels. Methods, which are concerned with the various aspects of assessment, are applicable to all levels of assessment. However, there is no universal set of methods that can be applied to all project types and in all impact assessment situations.

It is not possible to discuss all of the methods used in EIA in a single chapter. Thus, this chapter focuses on the suite of EIA methods most commonly used to assist impact identification, relative impact measurement, impact interpretation, and impact communication (see Baker and Rapaport 2005). Other examples of EIA methods and techniques are discussed at various places throughout this book, while methods and techniques for public involvement are discussed in Chapter 11.

METHODS FOR IMPACT IDENTIFICATION AND CLASSIFICATION

Methods for identifying impacts and issues are useful and systematic aids to the EIA process. Such methods alone do not constitute an EIA, as is often mistakenly believed; rather, they serve to organize and present information for further inquiry. One of the most important reasons for using such methods is that they provide a means for the synthesis of information and for the evaluation of alternatives on a common basis (Canter 1996). A variety of methods support impact and issues identification and organization of information in EIA. The choice of methods varies depending on the nature of the particular problem, the local socio-economic context, the availability of time and resources, and the goals and objectives of the assessment. No particular set of methods can do all that is required of impact assessment. Some of the more common methods for impact identification and classification are presented here, based on Bisset (1988), Canter (1992), Shopley and Fuggle (1984), and the UNEP (2002). Many of these methods will be revisited in other chapters.

Examination of Similar Projects

A starting point for EIAs is often the examination of similar projects, commonly referred to as **ad hoc approaches.** Ad hoc approaches are not really methods per se, since they do not structure the EIA problem, data, or information for the purpose of systematic identification of issues and impacts. Rather, ad hoc methods focus on identifying particular project design features, environmental impacts, and proposed management measures, as well as public reaction, from other similar projects and experiences. An example might be a team of experts visiting several different hydroelectric project sites in order to identify impact issues and concerns and relate them to the proposed project in question. Ad hoc approaches are valuable time-saving and cost-saving methods, particularly when multiple projects of a similar nature already exist in the proposed development area. Difficulties do emerge, however, when transferring lessons from one biophysical or socio-economic context to another, particularly with regard to ensuring that the project environments are sufficiently similar to justify a transfer of findings. Further, while ad hoc methods are useful for identifying broad areas of possible impacts, they offer minimal guidance for impact analysis. Regardless of their limitations, however, examination of similar projects or previous relevant experience is almost always a first step when evaluating proposed developments.

Checklists

Perhaps the simplest systematic method used in EIA is the checklist. **Checklists** are comprehensive lists of environmental effects or indicators of environmental impacts designed to stimulate thinking about the possible consequences of a proposed action. They are typically used in conjunction with other methods to ensure that a prescribed list of items, regulations, or potential actions and effects is considered in the EIA and to screen particular project actions (Box 4.1). Checklists are also commonly used to assist development proponents in complying with EIA system requirements. Sometimes called a screening checklist or initial environmental evaluation checklist, such checklists help a proponent to describe their project and determine whether it will be subject to an EIA and, if so, what type of assessment.

One approach is to use a **programmed-text checklist**, similar to a questionnaire, with a series of questions to be answered. For example, a typical question for a bridge construction project might ask, 'Is there a risk of riverbank erosion?' If the answer is 'yes', then the user may be directed to a subset of questions, such as 'Is the erosion likely to be severe enough to cause harm to fish habitat?' Programmed-text checklists are best suited for the kinds of projects that are frequently proposed and assessed, such as bridge construction, municipal land-use developments, and forest access road construction. Glasson, Therivel, and Chadwick (1999) propose three additional classifications of simple EIA checklists: **descriptive**, **questionnaire**, and **threshold of concern** (Box 4.2). The primary advantages of checklists are that they promote thinking about the range of potential issues and impacts emerging from a project in a systematic way and are relatively efficient and easy to use. Among the disadvantages, checklists:

- are typically not comprehensive of all potential impacts, and as a result not all impacts are considered;
- are often impractical to use if they are comprehensive;
- are often too general and not tailored to specific project environments;
- do not evaluate effects either quantitatively or qualitatively;

Box 4.1 Partial Checklist for a Bridge Construction Project

Proposed project activities:	
dredging	☑
blasting	☐
pier construction	☑
traffic diversion	☑
Affected physical components:	
water quantity	☐
water quality	☑
soil quality	☐
soil stability	☑
air quality	☐
Affected biological components	
fish populations	☑
spawning habitat	☑
cavity nesting bird habitat	☐
wildlife habitat	☐
rare or endangered species	☐
Affected socio-economic components	
employment	☑
noise	☑
health	☐

- are subjective and qualitative, meaning that different assessors may reach different conclusions using the same checklist;
- do not consider underlying environmental systems or cause-and-effect relationships and hence provide no conceptual understanding of impacts.

Box 4.2 Types of Simple EIA Checklists

Descriptive checklists: Provide guidance on how to assess certain impacts, including data requirements and potential information sources.

Data requirements	*Data or information source and techniques*
Water quality:	
water uses, baseline chemicals present, runoff data	water user surveys, water quality analysis, hydrological modelling
Employment impacts:	
economic base, workforce characteristics, job creation	industry survey, community profiling, regional multipliers

Questionnaire checklists: Propose a set of questions that must be answered when considering the potential effects of a proposed development.

Will the project cause pollution of air, water, or soil?	Yes	No	?
Will there be a discharge of solid or dissolved substances to waste water?	☐	☐	☐
Is there a risk of discharge of gases that are damaging to health or environment?	☐	☐	☐
Is there risk of a potential impact on drinking water?	☐	☐	☐
Will the activity cause discharge of dust to the atmosphere?	☐	☐	☐

Will the project cause waste problems?	Yes	No	?
Will waste be created during operations that is hazardous to human health?	☐	☐	☐
Is there a risk that tailings may contaminate local water and soil resources?	☐	☐	☐
Have the long-term environmental impacts of mine waste been considered?	☐	☐	☐
Is the proposed management of hazardous waste in compliance with standards?	☐	☐	☐

Source: Based on NORAD 1994.

Threshold of concern checklists: List environmental components that might be affected by the project actions, specific criteria for each component, and thresholds against which the project actions can be assessed.

Component	*Criterion*	*Threshold of concern*	*Action or alternative*
Human health	noise level	maximum 12 dB increase	______
Economics	benefit-cost ratio	2:1	______
Water supply	withdrawal rate	125,000 litres/day	______

Simple Matrices

While specific design and format often vary from project to project, **matrices** are essentially two-dimensional checklists that consist of project activities on one axis and potentially affected environmental components on the other (Figure 4.1). Matrices are perhaps the most commonly used method for impact identification and are particularly useful for identifying first-order cause-effect relationships between specific activities and impacts, as well as for providing a visual aid for impact summaries. The disadvantages of matrices are the same as those of checklists. EIA matrices can often be large and difficult to complete, and the volume of information can make them difficult to understand. There are several types of EIA matrices, but the two most basic types, upon which more sophisticated matrices are often developed, are **magnitude matrices** and **interaction matrices**.

Magnitude matrices. The magnitude matrix goes beyond simple impact identification to provide a summary of impacts according to their magnitude, importance,

Summary of Worst-Case Potential Impacts Prior to Mitigation							
Impact Rating – = No impact 0 = Negligible impact 1 = Minor impact 2 = Moderate impact 3 = Major impact **Project Component**	*Physical facilities*	*Atmospheric emissions*	*Liquid and solid releases*	*Noise*	*Lights and beacons*	*Additive impacts*	*Repetitive impacts*
Environmental Components							
Marine Plants							
Phytoplankton	0	–	0	–	–	0	–
Macrophytes	–	–	–	–	–	–	–
Microbiota							
Water column	0	0	1	–	–	1	–
Sediments	0	0	1	–	–	0/1	1
Zooplankton	0	–	0/1	–	–	1	–
Ichthyoplankton	0	–	1	–	–	1	–
Macrobenthos							
Hyperbenthos	1	–	0	–	–	0	–
Epibenthos	1	–	1	–	–	1	1
Biofouling Community	1	–	1	–	–	1	1
Fish and Commercial Shellfish							
Pelagics	1	–	0	–	0	0	–
Groundfish	0	–	0	–	–	0	–
Shellfish	0	–	1	–	0	1	1

Figure 4.1 Partial impact identification matrix from the Hibernia offshore oil development project.

or time frame (Glasson, Therivel, and Chadwick 1999). The best-known and most comprehensive magnitude matrix is the **Leopold matrix**, originally developed for the US Geological Survey by Leopold et al. (1971). The Leopold matrix consists of a grid of 100 possible project actions along a horizontal axis and 88 environmental considerations along a vertical axis, for a total of 8,800 possible first-order project–component interactions. Each cell of the matrix consists of two values: a quantification of the magnitude of the impact and a measure of impact significance. Where an impact is anticipated, the matrix cell is marked with a diagonal line in the appropriate row and column. The magnitude of the impact is then indicated within the top diagonal of each cell, typically on a scale from −10 to +10, and the importance of the impact or interaction is indicated in the bottom half of the diagonal (Box 4.3).

Box 4.3 Illustration of a Typical Section of the Leopold Matrix

Matrix instructions:

1. Identify all actions across the top that are part of the proposed project.
2. Under each action, place a diagonal slash in the cell at the intersection of each component on the side of the matrix where an impact is possible.
3. Indicate the magnitude of the impact with a value from 1 to 10 in the upper left of each cell, where 1 is a low and 10 is a high magnitude. Indicate + for a positive impact or − for a negative impact. In the lower right, indicate a value from 1 to 10 for the importance of the impact.

			Components and actions: modification of regime							
			a) exotic flora or fauna introduction	b) biological controls	c) modification of habitat	d) alteration of ground cover	e) alteration of groundwater hydrology	f) alteration of drainage	g) river control and flow modification	h) noise and vibration
A. CHEMICAL CHARACTERISTICS	1. Earth	a. mineral resources								
		b. construction material								
		c. soils								
		d. land form								
		e. force fields and radiation								
		f. unique features								
	2. Water	a. surface								
		b. ocean								
		c. underground								
		d. quality								
		e. temperature								

For example:

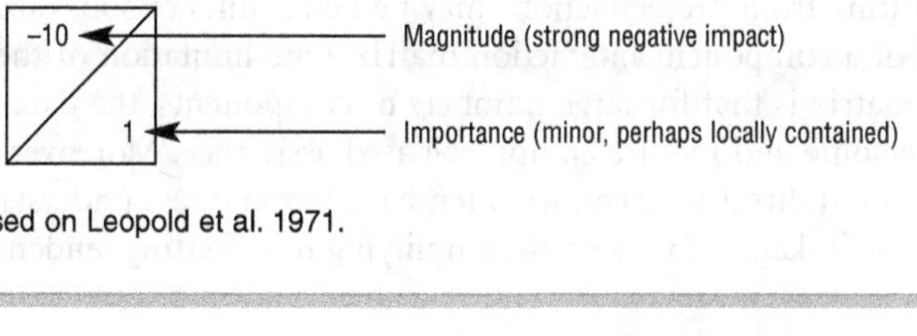

Source: Based on Leopold et al. 1971.

An advantage of the Leopold matrix is that it can easily be expanded or contracted based on the specific context of the project and environment being assessed. Although the Leopold matrix is perhaps the most comprehensive of EIA matrices, its sheer magnitude may be a drawback in itself. At the same time, it identifies only direct project impacts, has a primarily biophysical emphasis, and does not directly provide a framework for classifying the timing or duration of impacts. Given that the Leopold matrix was developed for use on many different types of projects, it tends to generate an unwieldy amount of information for any single project application (Glasson, Therivel, and Chadwick 1999). Moreover, it does not provide for direct weighting of the affected components to reflect their relative significance.

Weighted magnitude matrices. In an attempt to provide some indication of the 'relative importance' of identified project impacts, **weighted magnitude matrices** assign some measure of importance to each of the affected environmental components. In this way, values or judgments assigned to represent the potential impacts of a particular project action on an environmental component are multiplied by a weight to represent the relative importance of that component. The relative importance of the affected component may be a reflection of its current conditions, sensitivity to change, value to society, or importance in ecological functioning. Box 4.4 presents an example of a simple weighted magnitude matrix in which individual project action-component interactions are multiplied by the component weight and summed across the rows to determine the total impact on each component. Weighted magnitude matrices are particularly useful for comparing the relative impact of project actions across environmental components or for comparing alternative project locations. Specific techniques for assigning weights to environmental components include ranking, rating, and paired comparisons. These techniques are discussed in greater detail in Chapter 8. In the absence of a standardized system for assigning the impacts or the measures of component importance, magnitude matrices remain subjective and, like most other EIA methods, are only as good as the person assigning the values.

Interaction Matrices

A shortcoming of both the Leopold matrix and the weighted magnitude matrix is that neither goes beyond direct component interactions and impacts. Interaction matrices use the multiplicative properties of simple matrices to generate a quantitative impact score of the proposed project on interacting environmental components. There are two general types of interaction matrices: **component interaction matrices** and **weighted impact interaction matrices.**

First proposed by Environment Canada in 1974 (Harrop and Nixon 1999), component interaction matrices are intended to identify first-, second-, and higher-order interactions and dependencies between environmental components so that indirect impacts resulting from project actions may be better understood. Box 4.5 presents an example of a component interaction matrix. One limitation of the component interaction matrix is that for large numbers of components, the data may become quite cumbersome and require computer-based assistance. Moreover, as we reveal more and more indirect linkages, we often have less and less understanding of the nature of those linkages (for example, amplifying or offsetting tendencies) and less

Box 4.4 Example of a Simple Weighted Magnitude Matrix

Affected Environmental Components	Weight (importance)	Project actions: blasting	side cleaning	dredging	road construction	waste disposal	equipment transport	Total impact
air quality	0.26	–1			–1	–1		–0.78
water quantity	0.10	–2	–3	–3				–0.80
water quality	0.22	–2	–4	–2				–1.76*
noise	0.04	–2		–1	–2		–2	–0.28
habitat	0.08		–5		–3			–0.64
wildlife	0.08	–2	–4		–2			–0.64
human health	0.22	–2			+3	–3		–0.44

+ = positive impact
– = adverse impact

No impact =
Neglible impact = 1
Minor impact = 2

Moderate impact = 3
Major impact (irreversible or long-term) = 4
Severe impact (permanent) = 5

*Total impact (water quality) = (0.22)(–2) + (0.22)(–4) + (0.22)(–2) = –1.76

In the above matrix, the weights are distributed across the affected environmental components such that the total of all weights is '1', where the larger the weight the more important the component. In this way, all components can be given equal weight, 1/7 in the above matrix, but to increase the importance of one component requires that a trade-off be made and the importance of another component or components be decreased.

control over their impacts. Most impact management strategies can deal effectively with only primary and secondary impacts. This raises the question of how far down the impact chain we should go when identifying project-induced impact interactions. Is there an advantage to identifying potential third-, fourth-, fifth-, or higher-order linkages and interactions? Is it practical to do so?

Similar to weighted magnitude matrices in that impacts are multiplied by the relative importance of the affected environmental components, weighted impact interaction matrices explicitly incorporate second-order or indirect impacts. One of the more common examples of a weighted impact interaction matrix is the **Peterson matrix** (Peterson, Gemmel, and Shofer 1974) (Box 4.6). The Peterson matrix consists of three individual component matrices: a matrix that depicts the impacts of project actions or causal factors on environmental components; a matrix depicting the impacts of the resultant environmental change on the human environment; and a vector of weights or relative importances of those human components. The initial project–environment interaction matrix is multiplied by a matrix depicting secondary human–component impacts resulting from project-induced environmental change. The result is multiplied by the relative importance of each of the human

Box 4.5 Example of a Component Interaction Matrix for an Aquatic Environment

Step 1: Initial matrix of component linkages. Where a direct link exists between the dependent components on the side axis and the supporting components across the top, a value of '1' is entered in the cell; where there is no direct link, a '0' is entered. In other words, fish (row 4) are dependent upon vegetation (column 1), so a value of '1' is entered in the cell. Vegetation (row 1) is not directly dependent upon fish (column 4), so a value of '0' is entered.

		Supporting aquatic components				
Dependent aquatic components		vegetation	plant detritus	benthic fauna	fish	avifauna
	vegetation	0	1	0	0	0
	plant detritus	1	0	0	0	0
	benthic fauna	0	1	0	0	0
	fish	1	1	1	0	0
	avifauna	1	0	1	1	0

Step 2: Transfer all '1's' for direct impact values to a new matrix.

Step 3: Multiply the initial matrix by itself to solve for those cells that do not contain a '1' or direct link. If the result for any cell is greater than '0', then enter a value of '2' in the new matrix to indicate a 'second-order' link. If the result is zero, then enter a value of '0' in the matrix.

		Supporting aquatic components				
Dependent aquatic components		vegetation	plant detritus	benthic fauna	fish	avifauna
	vegetation	2	1	0	0	1
	plant detritus	1	2	0	0	0
	benthic fauna	2	1	0	0	0
	fish	1	1	1	0	0
	avifauna	1	2	1	1	0

The output is a component interaction matrix where a value of '1' indicates a direct link between aquatic components and a value of '2' suggests that the components are linked indirectly through another, common component.

Step 4: To identify possible third-order linkages, the procedure is repeated by multiplying the second matrix above by the initial matrix for all cells with no first- or second-order link to create a new matrix. If the result for any cell is greater than '0', then enter a value of '3' to indicate third-order component linkages, and transfer all '1s' and '2s' to this new matrix. To exhaust all possible linkages, the procedure can be repeated by multiplying the third matrix by the original, and so on.

components to generate an overall impact score. The advantage of the Peterson matrix lies in the multiplicative properties of matrices and the ease of manipulation. That said, its mathematical properties are also the primary limitation of the Peterson matrix in that two negative impacts, when multiplied, generate an overall positive impact. As discussed in the previous chapter, few impacts have offsetting tendencies.

Box 4.6 The Peterson Matrix for Route Selection

The following Peterson matrix was developed to score alternative routing options for a highway realignment project to bypass a local community. The project assessment scores for impacts on the biophysical environment were derived using expert judgment and based on experiences with similar highway realignment projects. The effects of the resulting biophysical change on the human environment, including the importance of the affected human components, were derived through community forums and focus groups in the potentially affected community.

The impact data are scaled in such a way that a 'minor' adverse impact is indexed as '1' and a 'major' adverse impact is indexed as '4'. The expert group classified a minor impact simply as one that can be fully mitigated with standard, known impact management actions and, based on previous experiences, is unlikely to cause long-term irreversible effects to the local biophysical environment. The expert group classified a major impact as one that was perceived as having a long-term social or economic effect on the livelihood or well-being of the community at large.

[A] Project routing options and affected biophysical effects

Project routing options	Affected biophysical VECS (VEC_b)				
	VEC_{b1}	VEC_{b2}	VEC_{b3}	VEC_{b4}	VEC_{b5}
(i)	3	2	4	1	4
(ii)	1	3	2	4	1
(iii)	3	2	4	3	2
(iv)	1	1	3	2	1
(v)	4	2	1	1	1

[B] Affected biophysical VECs and resulting effects on community

Affected biophysical VECS	Affected community VECS (VEC_c)				
	VEC_{c1}	VEC_{c2}	VEC_{c3}	VEC_{c4}	VEC_{c5}
VEC_{b1}	2	2	1	1	3
VEC_{b2}	1	1	2	2	3
VEC_{b3}	2	2	1	3	2
VEC_{b4}	1	2	4	3	2
VEC_{b5}	2	2	3	2	3

continued

[C] Importance (vector of weights) of community VECs

Community VECs		VEC weight
	VEC_{c1}	0.80
	VEC_{c2}	0.25
	VEC_{c3}	0.60
	VEC_{c4}	0.40
	VEC_{c5}	0.20

To calculate the Peterson matrix results, multiply the primary biophysical effects matrix [A] by the secondary community effects matrix [B], and multiply the resulting matrix by the vector of community VEC weights [C].

The result is an index for each project routing option as follows:

Project routing option:
i. 62
ii. 50
iii. 62
iv. 36
v. 37

The higher index is indicative of a higher overall effect and therefore a less preferred routing option. Each option can now be scaled or standardized to generate a relative ranking of routing options by using the following equation:

$$\frac{i_{max} - i}{i_{max} - i_{min}}$$

Where: i_{max} = the value of the routing option with the highest index
i_{min} = the value of the routing option with the lowest index
i = the value for each option, (i) through (v)

This result will yield a preferred option that is always 1, a least preferred option that is always 0, with all other scores falling somewhere in between. In this example, the scaled results are as follows:

i. 0.00
ii. 0.46
iii. 0.00
iv. 1.00
v. 0.96

Routing options (iv) and (v) are clearly the preferred options and more than twice as preferred as the next competing option (ii). Options (i) and (iii) are clearly the least preferred routing options.

Networks

Networks serve to identify potential direct and indirect impacts that may be triggered by initial project activities (Figure 4.2). Networks are useful for identifying sequential cause–effect linkages between project actions and multiple environmental

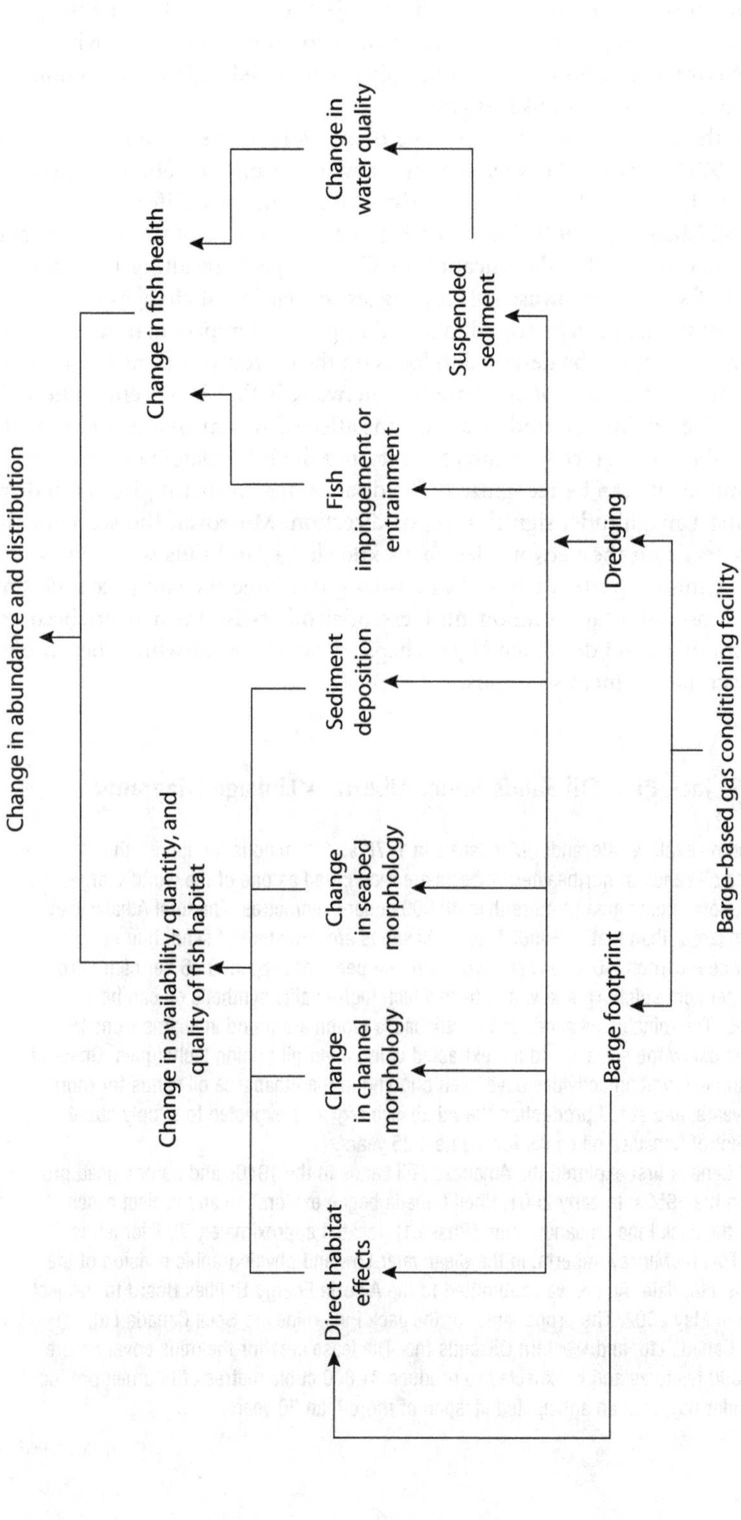

Figure 4.2 Effect pathways network diagram for the Niglintgak Barge Facility of the Mackenzie Gas Project, Northwest Territories.

Source: Adapted from the Mackenzie Gas Project Impact Statement, 2004, v. 5, section 7, Figure 7-8, p. 7-74.

components and can be used to describe how project activities could potentially lead to environmental changes that may affect certain components of the environment (Box 4.7). Networks can, however, become quite complicated and time-consuming as project actions and indirect linkages grow.

Perhaps the first and best known type of network is the **Sorensen network** (Sorensen 1971). A hybrid between a matrix and a network, the Sorensen network was initially developed to facilitate land-use planning in California (Glasson, Therivel, and Chadwick 1999). Box 4.8 depicts a partial section of the Sorensen network. The objective is the identification of direct impacts resulting from project actions and of subsequent cause–effect linkages so that initial changes in an environmental component can be traced to a final impact and impact management and monitoring schemes can be designed to focus on the correct components and linkages. A primary advantage of the Sorensen network is that it presents a **holistic approach** to identifying and understanding the affected system and its components. In this way, the importance of indirect effects and the interrelatedness of environmental components can be recognized. The disadvantage is that it gives no indication of impact magnitude, significance, or direction. Moreover, the scope of the network depends on the network designer's knowledge, and thus some important effects can be missed, particularly as the network grows large and complex. Inclusion of only the most important components is essential; otherwise, the network becomes too large, complex, and dominated by perhaps trivial effects for which there are no direct impact management solutions.

Box 4.7 Jack Pine Oil Sands Mine, Alberta—'Linkage Diagrams'

Described by explorer Alexander MacKenzie in 1778 as 'bituminous fountains', the Athabasca oil sands of northeastern Alberta are recognized as one of the world's largest oil deposits, covering an area of more than 40,000 square kilometres. The total Athabascan reserve is larger than that of Saudi Arabia. Oil sands are a mixture of sand, bitumen (83.2 per cent carbon, 10.4 per cent hydrogen, 9.4 per cent oxygen, 0.36 per cent nitrogen, 4.8 per cent sulphur), and water, from which high-quality synthetic oil can be processed. The minable oil sands in the Athabasca region are found in seams from 15 to 50 metres below the surface and are extracted using open-pit mining techniques. Open-pit mining and exploration activities have been ongoing in the Athabasca oil sands for more than 60 years, and at full production the Athabasca region is expected to supply about 10 per cent of Canada's oil needs for the next 25 years.

Shell Canada first explored the Athabasca oil sands in the 1940s and commenced production in the 1950s. In early 2001, Shell Canada began exploration and project assessment for the Jack Pine oil sands mine (Phase 1), located approximately 70 kilometres north of Fort McMurray, Alberta, in the Clearwater Lowland physiographic division of the Saskatchewan Plain. An EIS was submitted to the Alberta Energy Utilities Board for project approval in May 2002. The proponents for the Jack Pine mine are Shell Canada Ltd, Chevron Canada Ltd, and Western Oilsands Inc. The lease area for the mine covers more than 20,000 hectares and is expected to produce 31,800 cubic metres of bitumen product per calendar day, with an anticipated lifespan of more than 20 years.

continued

The major facilities of the proposed mine project include a 'truck and shovel' mining operation, a rock crusher and conveyor system, a processing plant, a tailings management system, and a co-generation plant, fuelled by natural gas, to produce power and steam. The purpose of Shell Canada's EIA was to examine the relationships between the activities associated with the mine development and potential impacts on the human and biophysical environment. To assist in identifying potential relationships, Shell Canada used a series of networks as 'linkage diagrams'. The linkage diagrams were used to describe how project activities would lead to environmental changes, which in turn could affect particular environmental components, and to identify key questions for impact investigation. The linkage diagrams consisted of four main components to represent project activities, potential changes in the environment, the central questions to consider, and critical links to or from other environmental or human components.

Given that the mine site itself is covered by a 1.5-metre-thick layer of muskeg and contains one open water body, a chief concern addressed in the Jack Pine EIA is effects associated with construction development and reclamation of the mine site on terrain units within the local study area. An impact linkage diagram (below) was developed to identify potential project effects, critical impact questions, and linkages to other environmental components.

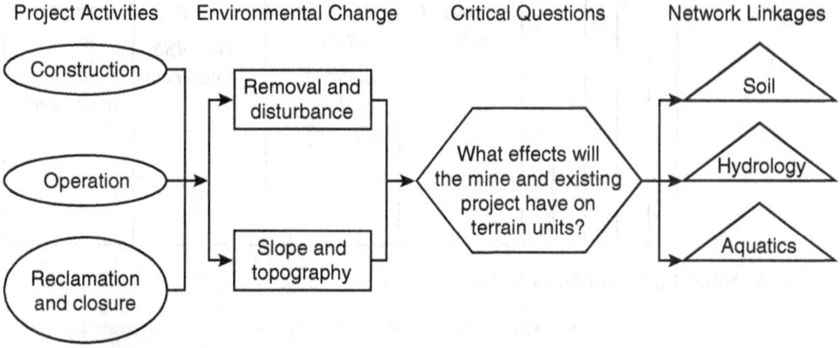

The predicted environmental impacts of the Jack Pine mine were deemed to be acceptable, generating no significant long-term impact on the environment provided that recommended mitigation measures are followed.

Source: Shell Canada 2002.

Overlays and Features Mapping

The use of overlays in impact mapping predates NEPA (McHarg 1968). Typically, transparencies depicting certain environmental or human components are overlain with project activities to identify areas of potential impact or concern (Figure 4.3). The components may be represented by their actual geographic area, such as would be the case for a protected space or housing area, or represented by map cells and assigned various colours or numerical values according to their importance or the significance of the impact. For example, a map may be created for the impacts of a particular project activity on the aquatic environment whereby the significance of a potential impact is identified by a numerical value '1' for a minor impact through '5'

Box 4.8 Partial Section of a Sorensen Network for Forestry

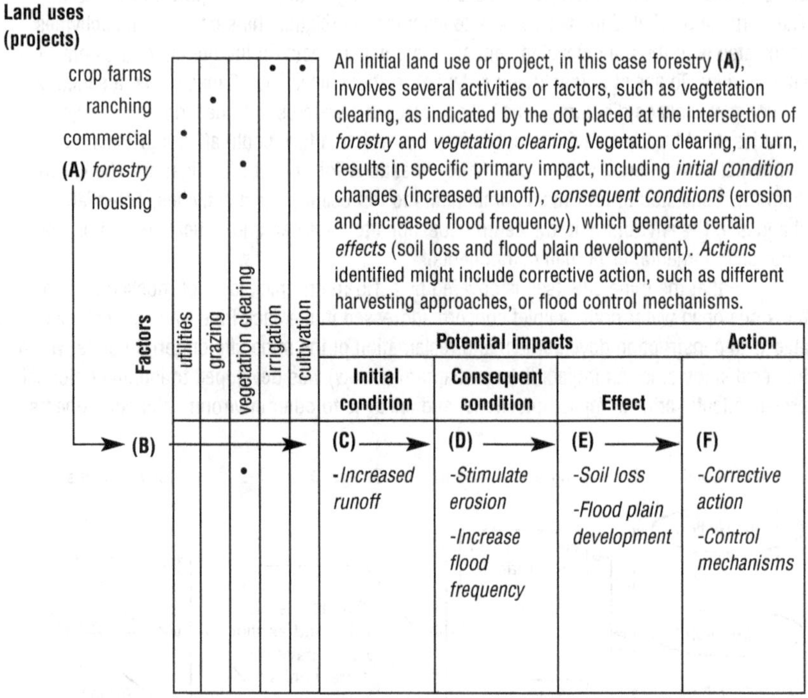
Land uses (projects)
crop farms
ranching
commercial
(A) forestry
housing
Factors
utilities
grazing
vegetation clearing
irrigation
cultivation
An initial land use or project, in this case forestry (A), involves several activities or factors, such as vegtetation clearing, as indicated by the dot placed at the intersection of forestry and vegetation clearing. Vegetation clearing, in turn, results in specific primary impact, including initial condition changes (increased runoff), consequent conditions (erosion and increased flood frequency), which generate certain effects (soil loss and flood plain development). Actions identified might include corrective action, such as different harvesting approaches, or flood control mechanisms.
Potential impacts
Action
Initial condition
Consequent condition
Effect
(B)
(C)
-Increased runoff
(D)
-Stimulate erosion
-Increase flood frequency
(E)
-Soil loss
-Flood plain development
(F)
-Corrective action
-Control mechanisms

Source: Adapted from Sorensen 1971.

for a significant impact. The same could be applied to other project activities and the maps overlain so that when the layers are summed, a high value indicates an area of a significant total impact. Such an approach is visually informative and allows spatial representation of impacts. Overlays and features mapping are also particularly useful when assessing land-use projects to identify potentially conflicting land-use patterns or optimal locations, such as the routing of an electrical transmission line or identification of marine shipping patterns so as to detect potential conflicts with fishing activities. The approach is limited, however, to only a small number of overlays because the resultant data quickly become difficult to manage and understand.

To address this limitation, the use of **Geographic Information Systems (GIS)** is becoming increasingly important in EIA. GIS are computer-based methods of recording, analyzing, combining, and displaying geographic information such as roads, streams, forest types, human settlement patterns, or any other feature that can be mapped on the ground. Although costly to operate and maintain, GIS are capable of handling very large data sets for each environmental parameter, and, effectively, an unlimited number of parameters and data points can be processed.

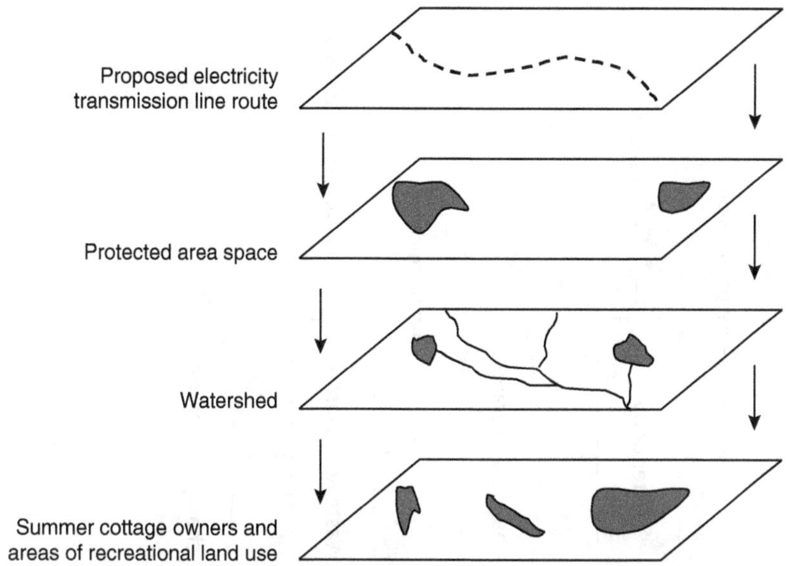

Figure 4.3 Illustration of environmental features mapping for impact identification.

Some of the limitations of traditional overlay systems, however, are still inherent to modern Geographic Information Systems. The likelihood and reversibility of impacts are not easily addressed; information for large impact areas is often too coarse to enable an effective identification of intricate interactions between environmental components; a large volume of data is required, and these data are currently not available for many remote and previously unexplored areas; the approach fails to effectively capture potential issues that cannot easily be mapped; and there is difficulty in identifying non-linear impacts.

Systems Models

Models are defined as simplifications of real-world **environmental systems**; they can range from box-and-arrow diagrams to sophisticated mathematical representation based on computer simulation. Systems models based on box-and-arrow diagrams are referred to as **systems diagrams**, which consist of environmental components linked by arrows indicative of the nature of energy flow or interaction between them (Figure 4.4). Bisset (1988) explains that systems diagrams are based on the assumption that energy flows and amounts can be used as a common unit to measure project impacts on environmental components. While such diagrams are useful, they serve more of a static function and are not well-suited to understanding dynamic environmental processes.

Systems models based on mathematical representation, on the other hand, are suited for meaningful study of dynamic environmental processes. A wide range of models is available, from those that deal with single issues such as air or water quality to complex ecological systems models. Examples include models to estimate the impacts of a project on air quality or simple demographic forecasting models.

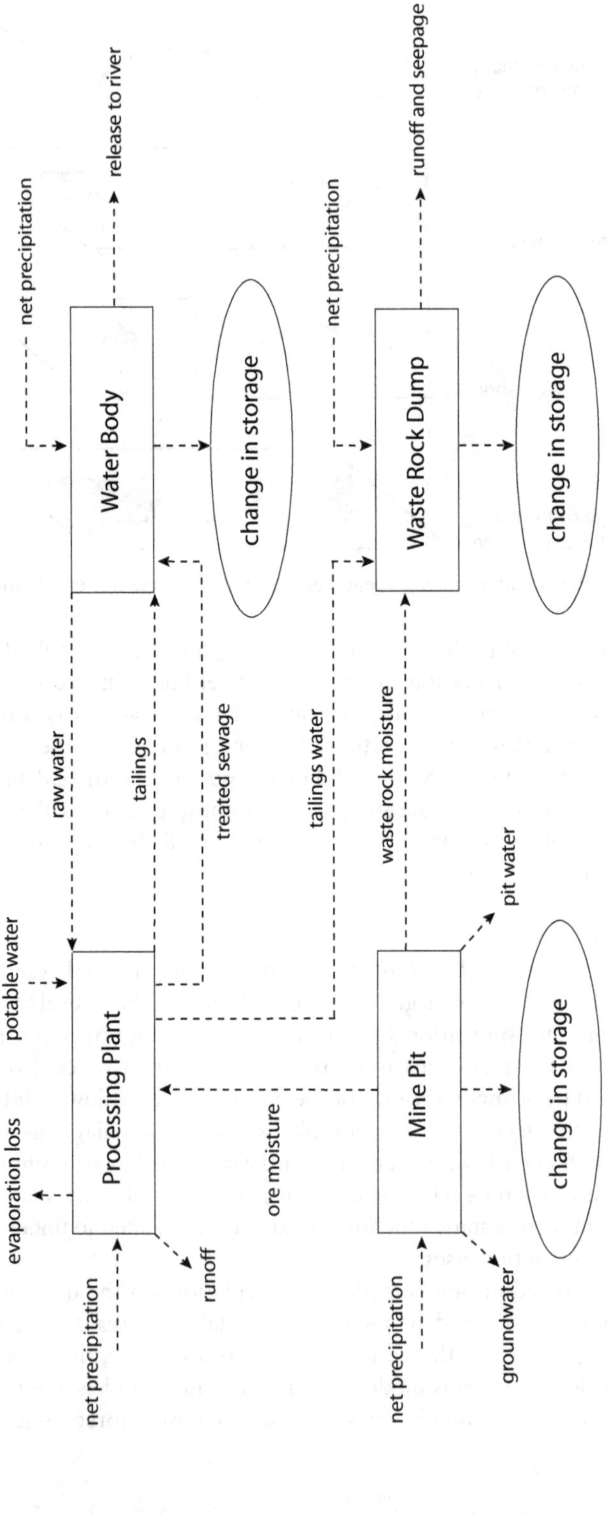

Figure 4.4 Systems diagram representing a mine site water balance.

Source: Based on BHPB 1998.

Perhaps the most popular, and earliest, modelling approach in EIA—**adaptive environmental assessment and management** (AEAM)—was introduced by C.S. Holling, C.J. Walters, and their associates at the University of British Columbia in the early 1970s. AEAM is an approach to simulation modelling for environmental problems that relies on teams of scientists, managers, and policy decision-makers working together to identify and define resource and environmental problems in quantifiable terms and then model alternative impacts and management solutions. While simulation modelling was the initial focus of AEAM, the scope has become much wider in recent years, and AEAM is currently applied as a broader conceptual approach to environmental assessment and management based on controlled experimentation and learning by doing (Noble 2004a).

Expert Judgment

While not exactly a 'method' in and of itself, expert judgment does underlie many other EIA methods, from issues and impact identification to impact prediction and management. As suggested by the United Nations Environment Programme (UNEP 2002), the successful application of many EIA methods relies heavily on the nature and quality of expert judgment. Bailey, Hobbs, and Saunders (1992), for example, reviewed artificial waterway developments in Western Australia and found that nearly two-thirds of all impact predictions were based on knowledge and experience. There is no standard by which expert judgment is integrated in EIA, and techniques for doing so range from unstructured and ad hoc, such as roundtable discussions, to highly structured and organized techniques such as the **Delphi technique** (see Chapter 7). As more complex methods involving integral reasoning become more common in EIA, there has been a correspondingly stronger reliance on expert judgment. Although there is no standard procedure for determining expertise, Noble (2004b) identifies several guidelines that have emerged from recent EIA practices (Box 4.9).

Three important points must be kept in mind regarding the use of expert judgment in EIA (Noble 2004b). First, no clear linear relationship exists between expertise and the 'quality' of impact prediction. Second, and contrary to common practice, consensus should never be the primary goal when relying on expert judgment, since the same lack of knowledge that required an EIA, which relies on expert judgment in the first place, is likely to guarantee that a group of diverse experts will disagree. Any consensus that is reached may be a forced consensus and therefore highly suspect. Thus, the dissenting perspectives become of particular importance for further exploration when relying on expert judgment. Third, and in light of the limitations of consensus, emphasis should be placed on consistency when relying on expert judgment (Box 4.10). **Consistency** simply refers to a measure of the extent to which expert judgments were purposefully made or reflect random decisions; in essence, it is a measure of the degree of understanding of the problem at hand.

METHODS SELECTION

There is no single best method that can accomplish all that is required for EIA, and each method has its own strengths and limitations (Box 4.11). Horberry (1984) examined

Box 4.9 Using Expert Judgment in EIA

Criteria for determining the desired number of experts and expert composition:

a) Sufficient representation of those affected by project and EIA decisions
 — affected publics and interest groups
 — affected sectors, government departments, and industries
b) Sufficient representation of those who affect project and EIA decisions
 — public administrators
 — planners and policy-makers
 — scientists and researchers
c) Appropriate geographic representation
d) Inclusion of necessary expertise and experience
e) Practicality, given time and resources
f) Large enough number to facilitate required techniques of data analysis
g) Credibility based on the number of expert views presented

Criteria for identifying and selecting experts:

a) Experience in two or more of the specialty areas considered in the assessment
b) Current or previous management or scientific leadership role in one or more of the specialty areas considered in the assessment
c) Experience in research or administration concerning one or more of the environmental or socio-economic components potentially affected by the project
d) Representation of a particular sector, interest, or affected geographic area
e) At least seven to 10 years of combined education and experience in EIA or in one or more of the key assessment areas (disciplines) involved
f) Experience in similar types of assessments or decision processes
g) A high level of professional productivity as evidenced by academic or professional publications and research, participation in academic or industry symposia, experience in project or environmental management, or previous membership on EIA decision or evaluation panels
h) Based on self-identified expertise—those who indicate expertise and wish to be involved

Source: Based on Noble 2004b.

140 EIAs and found 150 different methods being used! Multiple methods often are necessary in any single assessment. The objective is to select the methods best suited to the issue at hand while minimizing the overall limitations. When selecting methods for use in impact assessment, particular attention should be given to:

- *Objectives.* What is expected to be accomplished by method application?
- *Time availability.* How much time is available to collect data and apply the method?

Box 4.10 Expertise and Consistency

The notion that the judgment of experts is more consistent than that of non-experts seems almost tautological. However, a study by Noble (2004b) explored the relationship between expertise and consistency based on a case study of the impacts of electricity generation. Sixty-nine individuals participated in the study, and each was asked to self-identify his/her area of expertise based on the affected environmental components under consideration. Both experts and non-experts were then asked to provide their assessment of the impacts of various electricity supply alternatives on all environmental components. A consistency analysis of assessment judgments identified no significant statistical difference between the consistencies of expert versus non-expert assessments at the 95 per cent confidence interval. In fact, 60 per cent of judgments deemed to be 'extremely inconsistent' (close to random) were provided by those who self-identified themselves as experts on the particular component in question.

- *Resource availability.* What resources are available for method development and implementation, including human and financial resources?
- *Data availability.* What type and amount of data is required and available, including time series availability and availability for a range of environmental components?
- *Previous experiences.* Has the method been proven in previous applications?
- *Nature of the project.* Is the method appropriate given the size and scale of the assessment and its potential impacts?

According to Sadar (1996), desirable EIA methods should be:

- *comprehensive*—recognizing that 'environment' includes complex interrelationships between human and physical environments;
- *selective*—so that they can identify the impacts that are most important and critical to project decision-making;
- *comparative*—and therefore capable of differentiating incremental impacts likely to occur in the absence of the project from project-induced change;
- *objective*—and thus able to provide unbiased measures and information.

The best EIA methods are those that are capable of organizing large amounts of data and information, providing an overall summary of information in a way that is easily understood, allowing aggregation and disaggregation of data and information without losing information that is important to understanding and decision-making. The general rule of thumb is to use what is available, given the nature and quality of the data and the desired resolution, to get the job done.

Box 4.11 Comparison of EIA Methods

	1	2	3	4	5	6	7	8
Descriptive checklists	☐	☒	☒	☒	☐	☑	☑	☐
Questionnaire checklists	☐	☐	☐	☐	☐	☑	☑	☐
Threshold of concern checklists	☐	☐	☐	☐	☐	☑	☑	☐
Magnitude matrices (Leopold)	☒	☒	☒	☑	☑	☑	☑	☑
Weighted magnitude matrices	☑	☑	☒	☑	☑	☑	☑	☑
Component interaction matrices	☐	☒	☐	☐	☐	☐	☐	☐
Weighted impact interaction matrices	☑	☑	☑	☐	☑	☑	☑	☐
Network diagrams	☑	☒	☑	☒	☐	☒	☑	☒
Sorensen network	☑	☒	☑	☒	☐	☒	☑	☒
Overlays (GIS)	☐	☑	☐	☐	☐	☑	☒	☑
Systems models	☐	☐	☑	☐	☑	☒	☒	☑

1. Comprehensive of physical and human components.
2. Identifies relative importance of impacts or components.
3. Includes direct and indirect impacts.
4. Identifies positive and negative impacts, magnitude, likelihood, and reversibility.
5. Uses qualitative and quantitative data.
6. Summarizes key impacts for further consideration.
7. Cost-effective.
8. Capable of effectively communicating a large volume of useful information.

KEY TERMS

ad hoc approaches
adaptive environmental assessment and management (AEAM)
checklists
component interaction matrices
consistency
Delphi technique
descriptive checklist
environmental systems
Gaussian dispersion model
Geographic Information Systems (GIS)
holistic approach
interaction matrix
Leopold matrix
magnitude matrices
matrices
methods
models
networks
Peterson matrix
programmed-text checklist
questionnaire checklist
Sorensen network
systems diagrams
techniques
threshold of concern checklist
weighted impact interaction matrices
weighted magnitude matrices

STUDY QUESTIONS AND EXERCISES

1. In small groups, develop a simple questionnaire checklist to be used for assessing the construction phase of hydroelectric dam projects. Exchange your questionnaire checklist with other groups, and critically examine other groups' checklists. Are certain items not considered in the checklist that should be included? Compile an aggregate checklist of all groups' submissions, and discuss, among all groups, the advantages and limitations of a checklist approach to impact scoping.
2. Using Box 4.4 as a guide, construct, as a group, a simple matrix identifying project actions and affected environmental components for a proposed highway development. Once agreement is reached as to the project actions and affected components, break into small groups and assign weights and assessment scores to the components and impacts. Calculate the total impact score for each affected component, and sum the results to generate an overall project impact score. Compare your results to those of other groups. Are there noticeable differences in the results? Why might this be so? Discuss the advantages and limitations of the magnitude matrix method. How might these limitations be addressed?
3. Use the example presented in Box 4.5 to identify potential third- and fourth-order linkages between the dependent and supporting aquatic components. Discuss the usefulness and practicality of identifying linkages beyond the first and second orders for project and impact management.
4. Complete an example of the Peterson matrix identified in Box 4.6. Work in groups to identify and add project actions and affected components and to assign impact values on a scale from '1' to '5' where '1' indicates a minor impact and '5' indicates a major impact. Assign negative values where impacts are adverse. Assign weights to the human components on a scale of '1' to '5' where '1' indicates 'unimportant' and '5' indicates 'very important', and calculate the overall impact score. Discuss the usefulness of this value. What happened when the 'negative' impact scores were multiplied? Discuss the advantages and limitations of the Peterson matrix.
5. Sketch a simple network diagram for a proposed development that involves the clearing of a forested area. Provide a statement explaining the nature of each linkage in the network. Discuss the advantages and disadvantages of the network method.
6. Obtain a completed project EIS from your local library or government registry, or access one on-line. Scan the assessment and identify the types of methods used. Compare your results with those of others. Do certain methods appear to be common among most impact statements?

REFERENCES

Bailey, J., V. Hobbs, and A. Saunders. 1992. 'Environmental auditing: Artificial waterway developments in Western Australia'. *Journal of Environmental Management* 34: 1–13.

Baker, D., and E. Rapaport. 2005. 'The science of assessment: Identifying and predicting environmental impacts'. In K. Hanna, Ed., *Environmental Impact Assessment: Practice and Participation*. Toronto: Oxford University Press.

Barrow, C.J. 1997. *Environmental and Social Impact Assessment: An Introduction*. London: Arnold.

Bisset, R. 1988. 'Developments in EIA methods'. In P. Wathern, Ed., *Environmental Impact Assessment: Theory and Practice*, 47–61. London: Unwin Hyman.

BHPB (Broken Hill Proprieties Billiton). 1998. *NWT Diamonds Project Environmental Impact Statement*. Vancouver: BHP Diamonds Inc.

Canter, L. 1992. 'Advanced environmental impact assessment methods'. Paper presented at the 13th International Seminar on Environmental Assessment and Management, Centre for Environmental Management and Planning, Aberdeen.

———. 1996. *Environmental Impact Assessment*. 2nd edn. New York: McGraw-Hill.

Glasson, J., R. Therivel, and A. Chadwick. 1999. *Introduction to Environmental Impact Assessment: Principles and Procedures, Process, Practice and Prospects*. 2nd edn. London: University College London Press.

Harrop, D.O., and A.J. Nixon. 1999. *Environmental Assessment in Practice*. Routledge Environmental Management Series. London: Routledge.

Horberry, J. 1984. 'Development assistance and the environment: A question of accountability'. PhD thesis, MIT.

Leopold, L.B., et al. 1971. *A Procedure for Evaluating Environmental Impact*. Washington: United States Geological Survey, Geological Survey Circular no. 645.

McHarg, I. 1968. *Design with Nature*. Garden City, NY: Natural History Press.

Noble, B.F. 2004a. 'A state-of-practice survey of policy, plan, and program assessment in Canadian provinces'. *Environmental Impact Assessment Review* 24: 351–61.

———. 2004b. 'Applying adaptive environmental management'. In B. Mitchell, Ed., *Resource and Environmental Management in Canada*, 3rd edn, 442–66. Toronto: Oxford University Press.

NORAD (Norwegian Agency for Development Cooperation). 1994. *Initial Environmental Assessment: Mining and Extraction of Sand and Gravel*. Oslo: NORAD.

Peterson, G.L., R.S. Gemmel, and J.L. Shofer. 1974. 'Assessment of environmental impacts: Multiple disciplinary judgments of large-scale projects'. *Ekistics* 218: 23–30.

Sadar, H. 1996. *Environmental Impact Assessment*. 2nd edn. Ottawa: Carleton University Press.

Shell Canada. 2002. *Application for Approval of the Jack Pine Mine—Phase 1*. Environmental Impact Statement submitted to the Alberta Energy and Utilities Board and Alberta Environment. Edmonton.

Shopley, J., and R. Fuggle. 1984. 'A comprehensive review of current environmental impact assessment methods and techniques'. *Journal of Environmental Management* 6: 27–42.

Sorensen, J. 1971. 'A framework for identification and control of resource degradation and conflict in the multiple use of the Coastal Zone'. Berkeley, CA: Department of Architecture, University of California.

UNEP (United Nations Environment Programme). Economics and Trade Programme. 2002. *Environmental Impact Assessment Training Manual*. 2nd edn. New York: UNEP.

PART III

Environmental Impact Assessment Practice

CHAPTER 5

Screening Procedures

SCREENING

The number of projects that could potentially be subject to EIA is quite large; thus, the first stage in any EIA process is to determine whether the project requires an assessment and if so, to what extent. Screening is most often the responsibility of the project proponent, but in some cases the leading regulatory EIA authority may also play a role. **Screening** simply refers to the narrowing of the application of EIA to projects that require assessment because of perceived significant environmental effects or specific regulations. Essentially, screening is the 'trigger' for EIA and asks: 'Is an EIA required?' A screening process will result in one of the following decisions:

- no EIA is required;
- EIA is required;
- a limited EIA is required, consisting of a preliminary assessment or mitigation plan;
- further study is necessary, an initial environmental evaluation, to determine whether an EIA is required.

Screening ensures that no unnecessary assessments are carried out but that developments warranting assessment are not overlooked. However, as Barrow (1999) explains, screening caused problems during the first few years of the US NEPA. The primary screening procedure under NEPA was for 'major actions *significantly* affecting the environment', but *significantly* was an arbitrary concept—hence the importance of criteria and approaches for project screening.

SCREENING APPROACHES

Is an EIA Required?

There are ways to systematize the process of screening, thus improving accountability and decision transparency in determining whether a proposal requires an EIA. The most straightforward way is to compare the anticipated impacts of a project with the quality parameters of the environment specified in relevant legislation. Unfortunately, with the exception of Australia and California, listing quality parameters of the environment is not a common practice in EIA legislation. Requirements for screening are highly variable from one EIA system to another. Generally speaking, however, there are three approaches to EIA screening: case-by-case, threshold-based, and list-based.

Case-by-case screening, also referred to as discretionary or criterion-based screening, involves evaluating project characteristics against a checklist of regulations, criteria, or general guidelines as projects are submitted. In the province of Saskatchewan, for example, the need for an EIA is largely determined on the basis of screening criteria identified in the Saskatchewan Environmental Assessment Act. Section 2(d) of the Act indicates that any project, operation, or activity, or any alteration or expansion of such, is considered a development and subject to an assessment if it is likely to:

- have an effect on any unique, rare, or endangered feature of the environment;
- substantially utilize a provincial resource and pre-empt the use, or potential use, of that resource for other purposes;
- cause the emission of pollutants or create by-products or residual or waste products that require handling or disposal in a manner that is not regulated by any other Act or regulation;
- cause widespread public concern because of potential environmental changes;
- involve a new technology that is concerned with resource utilization and that may induce significant environmental change; *or*
- have a significant impact on the environment or necessitate a further development that is likely to have a significant impact on the environment.

Case-by-case screening such as under the Saskatchewan process allows for maximum flexibility to either *screen in* or *screen out* potential developments—it is sensitive to context, is dynamic, and provides for better consideration of the particular features of the local and regional environment in which development is proposed. At the same time, however, case-by-case screening can be time-consuming, inconsistent, and sometimes difficult to defend if the screening criteria are too vague.

Threshold-based screening involves placing proposed projects in categories and setting thresholds for each type, such as project size, level of emissions generated, or population affected (Box 5.1). Under threshold-based screening, thresholds are often based on different classes of development or project size or magnitude (e.g., total electrical generation capacity for a hydroelectric generating station) or based on environmental thresholds as established by regulations (e.g., total emission levels or concentrations). While threshold-based approaches are consistent and easy to use, a problem arises when projects lie just below the threshold. For example, if the threshold for a bridge construction project requiring an EIA is set at 50 metres, bridges that span 49 metres are not subject to EIA even though they are just as likely to generate similar environmental effects.

Thresholds for screening decisions are often part of listed requirements, or **list-based screening.** Also referred to as prescriptive screening, list-based screening involves a checklist of projects for which an EIA is (or is not) required, based on the potential of that project to generate significant effects or based on regulatory requirements. In California, for example, the California Environmental Quality Act lists projects for which a full EIA must always be completed, based on project characteristics, thresholds, and geographic location. As well, a negative list outlines projects for which an EIA is not required. Such lists are typically referred to as inclusive and exclusive project screening lists. **Inclusion lists** include projects that have either

Box 5.1 Threshold-Based Screening Regulations in Thailand

A 1978 amendment to the Improvement and Conservation of National Environmental Quality Act provided the legislative framework for EIA in Thailand, requiring that an EIA be carried out for certain types of projects and magnitudes of projects, including but not limited to:

- dams or reservoirs with a storage volume greater than 100 million cubic metres or storage surface area greater than 15 square kilometres;
- irrigation projects with a total irrigated area of more than 12,800 hectares;
- commercial ports and harbours with a capacity for vessels greater than 500 gross tonnes;
- thermal power plants with a generation capacity greater than 10 megawatts.

mandatory or discretionary requirements for EIA. **Exclusion lists**, on the other hand, include the sorts of projects that would be subject to an EIA unless they were on the exclusion list—for example, projects carried out in response to a national emergency, projects carried out for national security reasons, or routine projects considered of only minor significance.

The World Bank has explicit categories of projects that require assessment (Box 5.2) but cautions that project lists should be used flexibly with reference to particular geographic settings and project scale. The Nova Scotia Environmental Assessment Regulations 'Schedule A—Class I and Class II Undertakings' under section 49 of the Environment Act (1994–5, c. 1), for example, lists specific projects and thresholds for which an EIA is required—such as a facility for the incineration of municipal solid waste or a water reservoir with a storage capacity that exceeds the mean volume of the natural water body by 10,000,000 cubic metres or more. Similarly, Britain's EC Directive 85/337/EEC First Schedule, Part II 1989 Regulations, for example, identifies a list of the types of projects for which an EIA is required, some of which are defined according to certain thresholds.

Towards a Hybrid Screening Mechanism

List-based and threshold-based approaches to screening are relatively straightforward screening tools, are widely used, and provide for an efficient screening process. However, list-based approaches should be used cautiously and should not be used as the sole screening mechanism. The automatic requirement for assessment based on thresholds, for example, may result in circumvention of the assessment process when proponents propose developments that are just below the threshold. While such concerns are relevant to case-by-case screening as well, the important factor that is missing in list-based approaches is the opportunity to consider context—especially cumulative, compounding, or induced impacts. Additional problems associated with list-based screening, identified by Mayer et al. (2006) in an international study of screening mechanisms, include:

- listed project types, characteristics, or thresholds may be outdated and may not conform with the current state-of-the-art;
- thresholds may be identified even though there is no sound reason for them;

Box 5.2 World Bank Project Screening Lists

Category A Projects

These projects are likely to have significant adverse environmental impacts that are diverse, are unprecedented, or affect an area beyond the specific project site. A full EIA is required for such projects as:

- large-scale industrial plants
- dams and reservoirs
- port and harbour development
- large-scale irrigation
- river basin development
- hazardous or toxic materials involvement
- reclamation and new land development
- forestry and production projects
- large-scale land clearance
- oil, gas, and mineral development
- large-scale drainage or flood control
- thermal or hydropower development
- manufacture, transport, use of pesticides
- resettlement

Category B Projects

These projects are likely to have adverse impacts but are less significant than Category A projects. Most impacts are reversible, manageable, and site-specific. Projects include:

- electricity transmission
- renewable energy development
- tourism development
- small-scale irrigation and drainage
- rural water supply or sanitation
- watershed management or rehabilitation
- agro-industries
- rural electrification
- small-scale aquaculture
- rural electricity supply
- small project maintenance or upgrading

Category C Projects

These projects are likely to generate only minimal or no significant adverse environmental impacts. No EIA is required for projects concerning:

- education initiatives
- nutrition programming
- institution development
- most human resource projects
- family planning
- health initiatives
- technical assistance

Source: World Bank 1993.

- regional differences in the sensitivity of the environment are not considered;
- specified thresholds may not be appropriate in all situations and contexts;
- thresholds are set either too high or too low to properly capture significant impacts.

As summarized by Snell and Cowell (2006), lists have a useful role, but they cannot replace case-by-case considerations. In case-by-case or discretionary screening, determination of the need for assessment depends heavily on the relative weight given to each criterion considered and to the context within which significance decisions about the project's potential impacts are being made. In contrast to list-based approaches, case-by-case evaluations are dynamic and provide for a better consideration of particular features of the environment (e.g., environmental sensitivities, unique landscapes, endangered species) that are critical in significance determination. That said, case-by-case approaches are time-consuming and often inconsistent, and sometimes the decision is difficult to defend.

An alternative approach is to use the above screening mechanisms in combination such that projects above specified thresholds or located in sensitive areas, for example, are subject to mandatory assessment, whereas projects that fall below the threshold or not located in sensitive areas are screened on a case-by-case basis. A hybrid screening mechanism (Box 5.3) is a threshold-based screening system with allowances for case-by-case consideration. Under this approach, a prescriptive

Box 5.3 Hybrid Screening Approaches in the UK

EC Directive 85/337/EEC First Schedule, Part II of the 1989 Regulations is illustrative of *inclusion* and *threshold-based* screening, identifying projects and thresholds subject to EIA as including:

- urban development projects that would involve a total area greater than 50 hectares for new or extended urban areas and an area greater than two hectares within existing urban areas;
- oil and gas pipelines that exceed 80 kilometres in length;
- installations for the manufacture of cement;
- industrial estate development projects that exceed 15 hectares;
- waste-water treatment plants with a capacity greater than 10,000 population or equivalent;
- all fish meal and fish oil factories;
- petroleum extraction, excluding natural gas.

In addition, a *case-by-case* provision states that at the discretion of a local planning authority, an EIA may be required for projects smaller than the specified thresholds but likely to have a significant effect on the environment.

The nature of the hybrid approach is clearly illustrated by the DETR (1998) proposal, which includes an 'exclusive threshold' for projects that are exempt from EIA, an 'indicative threshold' for projects requiring case-by-case analyses, and an 'inclusive threshold' for projects for which EIA is always required.

EIA always required — **Inclusive threshold**

[*EIA more likely to be required, but test remains likelihood of significant environmental effect*]

Case-by-case consideration — **Indicative threshold**

[*EIA less likely to be required, but test remains likelihood of significant environmental effect*]

EIA not required — **Exclusive threshold** (except projects in sensitive areas)

Sources: Based on Gilpin 1999; EC Directive 85/337/EEC First Schedule Part II Regulations 1989; DETR 1998 (proposed amendment to Directive 97/11).

screening mechanism provides for a list-based screening approach to projects and thresholds for which assessment is always required, while a discretionary case-by-case screening mechanism is applied to projects that fall below mandatory thresholds, fall in between inclusion and exclusion thresholds or criteria, or are included in descriptive lists (DETR 1998).

SCREENING REQUIREMENTS IN CANADA

Requirements for EIA

The Canadian Environmental Assessment Act applies only to 'projects'. This includes:

- activities in relation to physical works;
- activities not related to physical works but described in the Inclusion List Regulations.

As noted in Chapter 2, the current federal EIA process, including the triggers for federal EIA in Canada, is currently under review. However, at the time this book was written, once any activity has been determined to be a project, as defined under the Canadian Environmental Assessment Act, then three additional criteria must be fulfilled before environmental assessment is applied at the federal level (Figure 5.1):

- the project must not be excluded from environmental assessment;
- a federal authority must be involved;
- there must be a trigger to initiate the assessment.

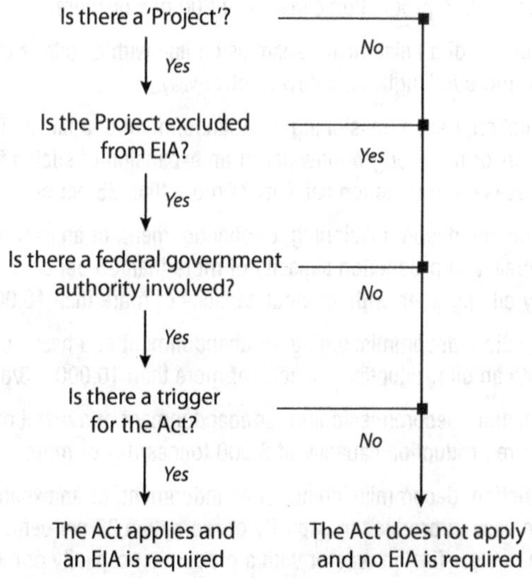

Figure 5.1 Determining the need for EIA under the Canadian Environmental Assessment Act.
Source: Canadian Environmental Assessment Agency. www.ceaa.gc.ca.

The Canadian federal Inclusion List Regulations specify activities not related to a physical work that may require an environmental assessment. These activities do not involve the construction of physical infrastructure and include the abandonment of the operations of a pipeline or mine site, the disposal or release of nuclear substances, the destruction of fish or fish habitat, activities proposed in a national park or national historic site, and projects on Aboriginal lands. In addition to EIA approval, physical activities and classes of physical activities set out under the Inclusion List Regulations typically require permission or approval under other federal acts and regulations such as the National Parks Act, the National Energy Board Act, Radiation Protection Regulations, the Navigable Waters Protection Act, the Indian Act, and Migratory Bird Regulations. The official Canadian Inclusion List Regulations can be accessed on the Department of Justice website at http://laws.justice.gc.ca/en/index.html.

Comprehensive Study List Regulations identify projects that are likely to cause significant adverse effects or are of significant public concern—for example, activities in a national park, nuclear facilities, and mining and mineral processing operations—for which an environmental assessment is required. Projects and activities for which an EIA is required under the Comprehensive Study List Regulations are typically categorized according to certain project thresholds (i.e., threshold-based screening), as discussed above and indicated in Box 5.4.

Box 5.4 Selected Screening Thresholds under the Canadian Federal Comprehensive Study List Regulations

The proposed construction, decommissioning, or abandonment of a fossil fuel–fired electrical generating station with a production capacity of 200 mw or more.

The proposed construction of an electrical transmission line with a voltage of 345 kV or more that is 75 km or more in length on a new right-of-way.

The proposed construction, decommissioning, or abandonment of a facility for the extraction of 200,000 m3/year or more of groundwater or an expansion of such a facility that would result in an increase in production capacity of more than 35 per cent.

The proposed construction, decommissioning, or abandonment, or an expansion that would result in an increase in production capacity of more than 35 per cent, of an oil refinery, including a heavy oil upgrader, with an input capacity of more than 10,000 m3/day.

The proposed construction, decommissioning, or abandonment of a heavy oil or oil sands processing facility with an oil production capacity of more than 10,000 m3/day.

The proposed construction, decommissioning, or abandonment of a metal mine, other than a gold mine, with an ore production capacity of 3,000 tonnes/day or more.

The proposed construction, decommissioning, or abandonment, or an expansion that would result in an increase in production capacity of more than 35 per cent, of a Class IA nuclear facility, i.e., a nuclear fission reactor with a production capacity of more than 25 mw (thermal).

The proposed construction of a railway line more than 32 km in length on a new right-of-way.

Certain projects are automatically excluded from the EIA process. This includes projects carried out in response to a national emergency and other activities described on the Exclusion List Regulations. Such projects and activities are deemed not likely to generate significant adverse effects, and in many cases they are similarly defined by screening-based thresholds. For example, the proposed expansion of an electrical transmission line by not more than 10 per cent of its current length and that would not be carried out beyond the existing right-of-way, involve the release of pollutants into a water body, or involve the placement of any structure in or near a water body is excluded from assessment under the Canadian Environmental Assessment Act.

If a project is subject to an assessment under one or more of the above regulations, the federal EIA process is triggered when a federal authority has decision-making responsibility in relation to the proposed project. This may happen when a federal authority either:

- proposes a project;
- provides some form of financial assistance to assist a proponent in carrying out a specified project;
- sells, leases, or transfers control or administration of federal lands for the purposes of allowing a project to be carried out; or
- provides a licence, permit, or approval for a project that involves federal Acts or regulations included in the Law List Regulations, such as the National Energy Board Act and the Canada Transportation Act, that enable a project to be carried out (see http://laws.justice.gc.ca/en/index.html).

In other circumstances, when the federal minister of environment receives a petition requesting that a particular project be assessed and the minister considers the project likely to cause significant adverse environmental effects across jurisdictional boundaries, then the minister has the authority to require an assessment even if the above regulatory conditions are not met.

Canada currently has a very different screening approach and a much less streamlined one than, for example, the list-based approach currently in effect for Commonwealth EIA in Australia. Although certain developments still require EIA and approval under state legislation in Australia, as do certain developments under Canadian provincial and territorial EIA legislation, Australian national or Commonwealth EIA, carried out under the Environmental Protection and Biodiversity Conservation Act, 1999, applies primarily to actions that will have a significant impact on 'matters of national environmental significance'. This is a much more streamlined process than the Canadian federal system, but at the same time, the state-based system in Australia also differs significantly from the Canadian provincial system. Under the Commonwealth, matters of national environmental significance include world heritage properties, Ramsar wetlands, listed threatened species and threatened ecological communities, listed migratory species, activities related to nuclear energy and uranium mining, the Commonwealth marine environment, and national heritage places. Excluded from Commonwealth EIA, regardless of the effects on matters of national significance, are such actions as certain forestry operations, the development of bioregional plans, conservation agreements, and strategic assessments.

Type of EIA Required

Not all projects must necessarily undergo the same level of assessment—regardless of the EIA system. For example, a proposed nuclear energy facility is likely to require a considerably larger and more complex assessment than a simple access road construction project. Determining the requirements for and level of EIA, however, is not always a straightforward process based solely on case-by-case, threshold, or list-based screening criteria. For example, the EIA procedures in Canada and Western Australia incorporate a two-tiered system for determining whether an EIA is required and if so, what type of EIA.

Under Western Australia's system, a notice of intent is submitted to the Australian Environmental Protection Agency describing the nature of the project, potential environmental effects, alternatives to the project, and proposed impact management measures. This is referred to as an **initial environmental examination** (IEE) or **environmental preview report** (EPR). This information is then used to determine whether an EIA is needed and what level of assessment is required. Based on Australia's procedures, the decision stemming from an IEE may be that:

- no further assessment is required;
- a full EIA is required;
- an examination by an independent commission of inquiry is necessary; or
- a less comprehensive public environmental report is required.

Under the current Canadian federal EIA system, once it is determined that an EIA is required, there are five assessment options, as described in the Canadian Environmental Assessment Act. They include **screening EIA**, **class screening EIA**, **comprehensive study EIA**, **review panel EIA**, and **mediation EIA** (Box 5.5).

Box 5.5 Types of Projects Subject to Different Classes of EIA in Canada

Screening EIA

Agriculture and Agri-Food Canada submitted a proposal in 2002 for a screening assessment of a proposal to remediate a contaminated site in Lacombe, Alberta. The project involved the operation of a mobile PCB destruction system to remove and dispose of wastes and contaminated soils.

Aquila Networks Canada, in cooperation with Parks Canada, submitted a model class screening report in 2003 for the ongoing operations and maintenance of power distribution facilities in Banff National Park. The screening report documents the current conditions of the environment near the distribution facilities, highlights the activities involved in the operation and routine maintenance of the facilities, and outlines the potential impacts and standard environmental management practices associated with such activities.

In 2004, Parks Canada submitted a replacement class screening report for special events in the Halifax Defence Complex, an area that includes five national historic sites. No single spe-

continued

cial event or activity is defined in the report; rather, potential impacts and management measures are documented for multiple types of activities and events that would normally occur at a national historic site, including concerts, sporting events, and other community events. Issues addressed in the report concern parking, fencing, noise control, and waste management.

Comprehensive Study EIA

Manitoba Hydro, in partnership with Nisichawayasihk Cree Nation, recently submitted a proposal to construct and operate a 200-megawatt hydroelectric generating facility, the Wuskwatim Hydro Generation Station. The proposed facility would be located on the Burntwood River in northern Manitoba. The comprehensive study report was released in 2005, summarizing the project, its potential environmental impacts, the results of public consultations, and proposed mitigation and follow-up plans. In 2006, the minister of environment announced that the project would not require assessment by a review panel in that the project was not likely to cause significant adverse environmental effects provided that the mitigation measures and follow-up program outlined in the comprehensive study report were implemented.

In 2002, the minister of environment approved the Deep Panuke offshore gas development project. PanCanadian Petroleum Limited submitted a proposal to develop the Deep Panuke offshore natural gas field, approximately 250 kilometres southeast of Halifax. The Deep Panuke gas field has estimated recoverable reserves of at least one trillion cubic feet of natural gas and 5 million barrels of liquid hydrocarbons. Based on the proposed mitigation measures, the project was considered not likely to cause significant adverse effects, and public concerns did not warrant referral to a review panel or mediator.

Review Panel EIA

In 1994, the minister of environment appointed an assessment panel to review BHP Diamond's proposal to develop and operate Canada's first diamond mine. The proposed project would involve open-pit and underground mining of five diamond-bearing kimberlite deposits at Lac de Gras, 300 kilometres northeast of Yellowknife. The review panel concluded that impacts of the project are predictable and mitigable. The project was approved in 1996, subject to certain conditions, and the panel made 29 recommendations to address environmental and socio-economic concerns.

In 2006, Newfoundland and Labrador Hydro submitted a project registration and description for the Lower Churchill Hydroelectric Generation Project. The project would involve the construction and operation of hydroelectric generating facilities and interconnecting transmission lines with the existing Labrador electricity grid. The project requires the construction of two dams and reservoirs, the largest of which is a 99-metre-high dam with a reservoir extending 225 kilometres and flooding an area of approximately 86 square kilometres. The project would result in an additional generating capacity of 2,800 megawatts for the province. An environmental assessment is required for the project under the Newfoundland and Labrador Environmental Protection Act. An EIA is also required at the federal level, under Section 5 of the Canadian Environmental Assessment Act, because Fisheries and Oceans Canada is the responsible authority for issuing a permit or licence for development under the Fisheries Act and the project also requires approval under the Navigable Waters Protection Act. In 2007, the federal minister of environment announced that the project would undergo a federal environmental assessment by an independent review panel. This will be a joint review panel under both the Environmental Protection Act and the Canadian Environmental Assessment Act. The proponent is currently preparing the EIS for the project.

Screening EIA. As of early 2009, the majority of EIAs completed in Canada at the federal level are screening assessments. A screening EIA systematically documents the potential environmental effects of a proposed project, identifies mitigation measures, and determines the need to modify the project plan or recommends further assessment to eliminate or minimize those effects. A screening EIA must address:

- the environmental effects of the project and the effects of possible accidents;
- the significance of the environmental effects;
- technically and economically feasible measures that would reduce or eliminate any significant adverse effects;
- public comments.

If the project is likely to cause significant adverse environmental effects and it is uncertain whether these effects are justified in the circumstances, or public concerns warrant it, the project is referred to a mediator or a review panel. If a screening EIA determines that the significance of project impacts are acceptable and manageable given the circumstances, then action can be taken to allow the proposed project to proceed to the development phase. Current and archived screening EIAs can be searched at www.ceaa.gc.ca.

Class screening EIA. As a particular type of screening assessment, class screening EIAs are associated with routine undertakings, such as those involved in urban building projects or residential zoning. Class screening EIA helps to streamline the EIA process for certain types of projects not likely to cause significant adverse environmental effects, provided that certain design standards and mitigation measures are implemented. **Model class screenings** are used to provide a generic assessment of all projects within a certain class. Information contained in a 'model' report is applied to individual projects and adapted for location-specific or project-specific information. Such screenings might include fish habitat restoration programs or the construction of forest access roads. Similarly, **replacement class screenings** provide a generic assessment of all projects within a class, but no location- or project-specific information is required, and thus a screening report is not necessary for each individual project. An example would be a screening EIA prepared for special holiday events or activities in a national park or at a national historic site. Activities in a national park require assessment under the Inclusion List. They are also part of a well-defined class of activities, occur in well-understood environmental settings, are unlikely to cause significant adverse effects, do not require monitoring measures, involve straightforward planning and decision-making, and are unlikely to generate public concern. In such cases, a replacement class screening would be prepared that sufficiently addresses the nature of activities and special events in general, such as vehicle parking arrangements and management measures or the use of fireworks, but individual events need not undergo separate assessments. Examples of class screenings under the Canadian Environmental Assessment Act can be found at www.ceaa.gc.ca.

Comprehensive study EIA. Most federal projects are assessed through screening EIA. However, projects described under the Comprehensive Study List Regulations, typically associated with large-scale, complex, and environmentally sensitive projects with greater potential for adverse environmental effects, require a comprehensive EIA. A comprehensive EIA must address the same factors as a screening EIA as well as:

- a detailed description of the project;
- alternative means of carrying out the project that are technically and economically feasible and that consider the environmental effects of the alternative means;
- the capacity of renewable resources likely to be affected;
- the need for and any requirements of a follow-up program.

Projects such as large-scale energy projects, mine development projects, and industrial plants typically undergo comprehensive EIAs. Such projects may also be referred to a mediator or panel review assessment if deemed appropriate by the federal minister of environment. For comprehensive EIAs, the minister can request that a proponent submit additional information concerning project effects management and monitoring or require that the proponent address specific public concerns prior to receiving project approval.

Review panel EIA. A project assessment under review panel EIA is required if the environmental effects of a project are uncertain or potentially significant, if public concern warrants it, or if the minister of environment states that a review by an independent panel is required for either a current screening or a comprehensive study EIA. Review panels are appointed by the federal minister of environment.

Mediation EIA. As a voluntary process of negotiation, a mediation EIA involves an independent mediator to help interested parties resolve their issues. Mediation EIA can be used to address all issues that arise in an EIA, or it can be used in combination with an assessment by a review panel. Few mediation EIAs have been carried out in Canada.

SOME GENERAL GUIDELINES FOR SCREENING

Barrow (1999) suggests that screening criteria for determining the need for an EIA might include, at least in principle, the following:

- Some component of the project will likely reach a specified threshold.
- The site for which development is proposed is sensitive.
- The proposed development involves known or suspected dangers or risks.
- The development will potentially contribute to cumulative impacts.
- There are unattractive input-output considerations, such as excessive labour migration or pollution output.

Based on Annex 1 of the European EIA Directive, a number of general criteria are used to determine the need for some level of EIA:

- general condition and character of the receiving environment;
- potential impacts of the proposed development;
- resilience of the affected physical and human systems to cope with potential project-induced change;

- level of confidence associated with likely project impacts;
- consistency or compatibility with existing policy or planning frameworks;
- the degree of public interest or perceived effects.

Screening plays a critical role in the EIA process; it is that part of the process when a decision is made as to whether a project requires an EIA and what level of assessment is necessary. While countries with EIA systems are starting to adopt national and formal requirements and guidelines for screening, Barrow (1999) suggests that screening is not widely practised and that in cases where screening is done, authorities often fail to screen adequately. The result is that many projects that may have significant adverse effects are often overlooked and thus proceed with little prior knowledge and understanding of their potential environmental consequences. In this regard, a more precautionary approach to screening may be warranted.

Screening and the Precautionary Principle

EIA can be seen as fostering precaution insofar as it provides a mechanism for anticipating adverse environmental impacts associated with a proposed development (Glasson, Therivel, and Chadwick 1999). Within the context of screening, precaution focuses on determining what is a *sufficiently* sound or credible basis for requiring an EIA. The screening stage is pivotal in this regard in that it ensures that activities that are 'likely' to cause significant environmental impacts are given proper treatment and assessment. The problem is that it is often not possible at the screening stage for the competent authority to prove, with scientific certainty, the significance or insignificance of the impacts of a project. The suggested approach is to err on the side of caution; however, as Snell and Cowell (2006) explain, avoiding harm in the absence of evidence does not always sit well with the ethos of EIA in that decisions should be made on the basis of sound information.

The **precautionary principle** suggests that when scientific information is incomplete but there is a threat of adverse impacts, the lack of full certainty should not be used as a reason to preclude or to postpone actions to prevent harm. In other words, when considerable uncertainty exists as to whether a proposed activity is likely to cause adverse environmental effects, the lack of certainty should not be a reason for approving the proposed activity, for not requiring an EIA, or for not requiring rigorous mitigation and monitoring measures (IAIA 2003). Erring on the side of caution, the precautionary principle as a screening guideline places the burden of proof on the proponent, making the proponent responsible for demonstrating 'insignificance' (Lawrence 2005).

However, the precautionary principle is not without problems, and the principle itself has been misused to justify everything from minimal change to rejecting any project proposal. Moreover, the interpretation of precaution within EIA is troubled by concerns over the efficiency of the process as well as by the need to ensure that precaution about likely, significant impacts does not unduly constrain development (Snell and Cowell 2006). That said, the underlying objective of adopting a precautionary approach to screening is to provide better assurance that potentially significant effects are captured during the screening process and that an EIA is carried out when needed.

KEY TERMS

case-by-case screening
class screening EIA
comprehensive study EIA
environmental preview report
exclusion list
inclusion list
initial environmental examination
list-based screening
mediation EIA
model class screenings
precautionary principle
replacement class screenings
review panel EIA
screening
screening EIA
threshold-based screening

STUDY QUESTIONS AND EXERCISES

1. What is the purpose of screening in EIA? Should *all* proposed developments be subject to EIA?
2. Discuss the relative advantages and disadvantages of case-by-case, threshold-based, and list-based screening approaches. Which type of screening approach is used in your political jurisdiction (e.g., province)?
3. What types of projects or developments are subject to EIA in your country? Compare this to one or more of the national EIA systems listed in Table 1.3.
4. Visit the Canadian federal EIA registry at www.ceaa.gc.ca . Search the registry based on keywords such as 'mine', 'road', 'electricity' or by province or date. Identify one current or completed EIA for each of model screening, class screening, comprehensive study, and panel review. Provide a brief summary of the proposed project, including the proponent, location, and date submitted and approved—if available.

REFERENCES

Barrow, C.J. 1999. *Environmental Management: Principles and Practice.* New York: Routledge.

DETR (UK Department of Environment, Transport and the Regions). 1998. *Draft Town and Country Planning (Assessment of Environmental Effects) Regulations.* London: DETR.

Gilpin, A. 1999. *Environmental Impact Assessment.* Cambridge: Cambridge University Press.

Glasson, J., R. Therivel, and A. Chadwick. 1999. *Introduction to Environmental Impact Assessment.* London: Spon Press.

IAIA (International Association for Impact Assessment). 2003. 'Social impact assessment international principles'. Special Publication Series no. 2. Fargo, ND: IAIA.

Lawrence, D. 2005. 'Significance criteria determination in sustainability-based environmental assessment'. Report to the Mackenzie Gas Project Joint Review Panel. Langley, BC: Lawrence Environmental.

Mayer, S., et al. 2006. *Projects Subject to EIA: D 2.4 Report WP 4.* Vienna: Austrian Institute for Regional Studies and Spatial Planning.

Snell, T., and R. Cowell. 2006. 'Scoping in environmental impact assessment: Balancing precaution and efficiency?' *Environmental Impact Assessment Review* 26: 359–76.

World Bank. 1993. 'Sectoral environmental assessment'. *Environmental Assessment Sourcebook Update* no. 4.

CHAPTER 6

Scoping and Environmental Baseline Assessment

EIA SCOPING

As discussed in Chapter 1, EIAs conducted following NEPA and in the early years of the Canadian Environmental Assessment Review Process (EARP) were often unfocused, full of technical details, voluminous, difficult for the public to understand, and thus of little use for timely direction of environmental decision-making. It is now recognized that to consider all potential project impacts, environmental components, and issues in any single EIA is neither feasible nor desirable. **Scoping** determines the important issues and parameters that should be addressed in an environmental assessment, establishes the spatial and temporal boundaries of the assessment, and focuses the assessment on the relevant issues and concerns. It identifies the components of the biophysical and human environment that may be affected by development and for which there is scientific and public concern. This involves determining what elements of the project to assess, what environmental components are likely to be affected, how these environmental components have changed over time and what factors have driven such change, and how these components may be affected by other actions or disturbances in the project environment.

SCOPING TYPES AND FUNCTIONS

There are two broad types of scoping: closed and open. In **closed scoping**, the content and scope of the EIA are predetermined by law. Any modifications occur through closed consultations between the proponent and the competent authority or regulatory agency. In **open scoping**, the content and scope of the EIA are determined by a transparent process based on consultations with various interests and publics.

No matter what the type of scoping, the scoping process serves a number of important functions in the EIA process:

- ensuring input from potential stakeholders early in the process;
- identifying public and scientific concerns and values;
- evaluating these concerns to focus the assessment and provide a coherent view of the issues;
- ensuring that key issues are identified and given an appropriate degree of attention;
- reducing the volume of unnecessarily comprehensive data and information;

- avoiding a standard inventory format for EIA that may miss key elements or issues;
- defining the spatial, temporal, and other boundaries and limits of the assessment;
- ensuring that the EIA is designed to maximize information quality for decision-making purposes.

SCOPING REQUISITES

Similar to project screening, scoping requirements and practices vary considerably from one EIA system to the next. Canada and the Netherlands, for example, have a formal scoping stage as part of a legislated EIA process. In New Zealand, there are specific objectives for scoping processes (Box 6.1). Under UK Directive 85/337, scoping is interpreted as in part mandatory and in part discretionary. Under the amended UK Directive 11/97, for example, the competent authority, at the request of the developer, is required to give opinions as to what should be addressed in EIA, but how scoping is carried out in terms of the consideration of alternatives or identification of indirect impacts is largely at the discretion of the proponent.

While the requirements and guidelines for scoping vary, a number of general principles should be observed to facilitate good-practice EIA scoping:

- describe the proposed activity;
- scope project alternatives;
- identify valued ecosystem components;
- delineate the assessment spatial and temporal boundaries;
- establish the environmental baseline and trends;
- identify potential impacts and issues of concern.

Describe the Proposed Activity

The project description serves two purposes: first, to determine the need for an environmental assessment; second, to facilitate efficiency and co-ordination of the environmental assessment. Thus, description of the proposed activity for EIA scoping

Box 6.1 New Zealand Ministry for the Environment Scoping Objectives

- Identify potential impacts of proposed development on the environment.
- Identify potential impacts of subsequent environmental change on society.
- Inform potentially affected people of the proposed project.
- Understand the values of individuals and groups potentially affected by the project.
- Evaluate concerns expressed and the potential environmental impacts.
- Define boundaries for any further required assessment.
- Determine the nature of any further required assessment, including analytical tools and consultation methods.
- Organize, focus, and communicate findings, potential impacts, and concerns.

Source: New Zealand Ministry for the Environment 1992.

must be presented at a level of detail significantly greater than at the screening stage. Under the Canadian Environmental Assessment Act, for example, three tiers of project information must be provided, including:

1. General information:
   - nature, name, and proposed location of the project;
   - information on consultation with affected parties and concerned authorities;
   - details on other provincial, territorial, and federal EIA legislation or land claims agreements to which the project might be subject;
   - information concerning the proponent and the nature of the industry or organization;
   - identification of affected or involved government departments and agencies, including landownership;
   - information concerning required permits and project authorizations.
2. Project information:
   - project component structures, including size and capacities;
   - activities and activity scheduling associated with construction, operation, and decommissioning phases;
   - scheduling, engineering design specifications, and land-use patterns;
   - resources and material requirements, consumption volumes, and use patterns;
   - waste production, discharges, and management.
3. Site-specific information regarding the project:
   - project location and extent of operations;
   - location and distribution of potentially affected environmental components;
   - identification of likely affected environmental components;
   - description of previous and current land-use patterns and impacts.

Project information is normally provided by the project proponent; however, other valuable sources of project information, such as previous land-use patterns and conditions, can be obtained through consultations with the general public, professionals, governments, and special interest groups and can be based as well on experiences elsewhere.

'Need for' and 'purpose of' the project. Describing the 'need for' and 'purpose of' the proposed project is important for defining the project effectively. The 'need for' a proposed project is the particular problem or opportunity that the project is intending to address or satisfy. The 'purpose of' the project is what is intended to be achieved by actually carrying out the project. The need for and purpose of the project are typically established from the perspective of the project proponent. For example, the need for a proposed electrical generation station may be defined on the basis of the demand for electricity supply or the inability of the current system to supply reliable electricity. The purpose of the project might be to supply cost-efficient, reliable electricity to expanding city residential neighbourhoods in a manner that is profitable to the utility company. The need for a particular project and its purpose are usually established from the perspective of the project proponent and set an important context for the identification and evaluation of project alternatives.

Scope Project Alternatives

Many environmental impacts can be prevented before irrevocable project location and design decisions are made by considering options to those proposed. The consideration of alternatives is a central element to good-practice EIA and is described by the US Council on Environmental Quality as the heart of the EIA process. Alternatives are defined as options or different courses of action to accomplish a defined end. There are two types of alternatives that should be considered—alternatives to the project and alternative means of carrying out the project (Figure 6.1).

Alternatives to a project are functionally different ways of meeting the need and purpose of the described project. Proposing alternatives to a proposed project involves developing criteria and objectives for environmental, socio-economic, and technical variables and identifying the preferred alternative based on the relative benefits and costs of each option with reference to the identified variables.

In practice, 'alternatives to' are rarely considered comprehensively in project EIA. When they are considered, they often reflect narrow agency goals or are inherently biased towards the proposed project. Alternatives to a proposed project are typically limited to the 'no action' alternative, which is interpreted as either 'no change' from the current or ongoing activity or 'no activity', which refers to not proceeding with the proposed development. This bias, to a certain extent, is understandable in that by the time a project is proposed, a proponent has already dedicated considerable resources to it. For example, land for the construction of an industrial complex may already have been leased by the time a proponent submits an application for development. Many alternatives that may be more environmentally sound or socially beneficial have thus already been foreclosed by the time the EIA commences. Arguably, 'alternatives to' are best addressed at the early planning stages through the process of strategic environmental assessment (see Chapter 13) rather than by the proponent at the stage of project proposal (Box 6.2).

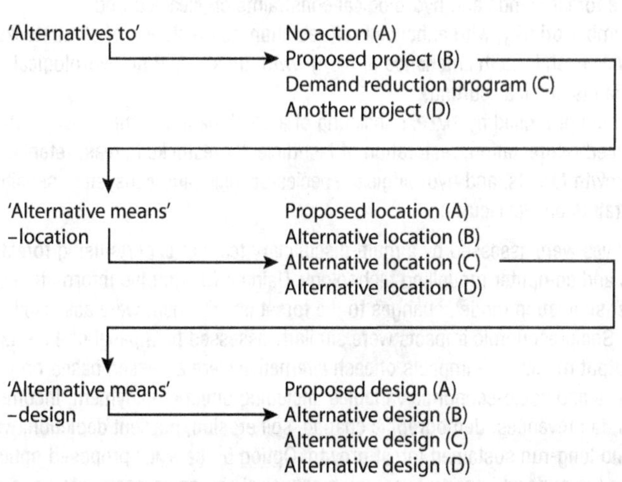

Figure 6.1 Hierarchy of project alternatives.

Box 6.2 Pasquia-Porcupine Forest Management Plan Alternatives Assessment

In 1995, a forest harvesting and management partnership was formed between MacMillan Bloedel Limited, one of Canada's largest forestry companies, and a subsidiary of the Saskatchewan Crown Investments Corporation, together known as the Saskfor-MacMillan Limited Partnership (SMLP). An application for development of the Pasquia-Porcupine forest was forwarded to the Saskatchewan government for approval. The Pasquia-Porcupine forest management area is located along the Saskatchewan–Manitoba provincial border within the Boreal Plain Ecozone. It encompasses an area of approximately 2 million hectares, of which more than half is suitable for commercial timber production.

The environmental assessment and forest management plan were endorsed in 1997. Consistent with the broader vision of sustainable forest management and based on early public consultation, eight plan objectives were proposed for the forest management plan:

- to provide quality products to meet customers' needs and provide a fair return to stakeholders;
- to provide safe and stable jobs;
- to safeguard heritage resources and traditional uses of the forest management area;
- to maintain the diversity of life forms in the area, including species and ecosystems;
- to maintain and enhance the forest ecosystem's long-term health;
- to minimize hazards from forest fires, insect infestations, and diseases;
- to protect primary resources of air, water, and soil;
- to ensure that forest areas regenerate after harvesting.

Based on these objectives, five alternatives were considered in the assessment:

1. no timber harvesting (the current baseline condition);
2. low timber priority, with reforestation based on historical levels, including retention of mature forest stands and hydrological constraints on clear-cutting;
3. intermediate timber priority, with enhanced reforestation, including retention of mature forest stands and hydrological constraints on clear-cutting;
4. high timber priority, with enhanced reforestation, reforestation of all unstocked productive land, no maintenance of old-growth forests, and no hydrological constraints on clear-cutting;
5. the option proposed by SMLP, consisting of a combination of the above, with enhanced reforestation, restoration of insufficiently restocked areas, retention of old-growth forests, and hydrological, species-specific, and ecosystem sensitivity constraints on clear-cutting.

Alternatives were assessed by a multi-disciplinary team of experts using forest land inventories and computer modelling technology. Using a Geographic Information System and a forest simulation model, changes to the forest environment were assessed for each alternative. Socio-economic impacts were similarly assessed by a panel of experts, using an input-output model. The impacts of each alternative were assessed based on a variety of biophysical and socio-economic variables, including direct employment, income projections, tax revenues, demographic change, soil erosion, nutrient depletion, water turbidity, and long-run sustained forest growth. Option 5, the SMLP proposed option, was identified as the preferred option based on biophysical and socio-economic trade-offs.

Sources: Noble 2004; SMLP 1997.

Alternative means can effectively be addressed at the project scoping stage. Alternative means refers to different options for carrying out a project—it has been accepted at this stage that the proposed project is the most suitable alternative. Alternative means typically include alternative ways a project can be implemented, such as technical or engineering design or alternative locations. However, it is not always possible to consider alternative locations for all proposed developments. For example, there are few possible alternative locations for a proposed gold-mining operation other than where the mineral deposit is located. That said, alternative access roads to the mine site might avoid ecologically sensitive habitats, and alternative designs for the access road can be identified to minimize collisions with wildlife.

Alternative means should be feasible in terms of meeting the purpose of the project and can be identified through previous or similar experiences elsewhere, expert judgment, public consultation and brainstorming, and/or more complex decision support systems. The number of alternative means can be narrowed by eliminating those that fail to meet certain minimum project, environmental, or socio-economic requirements. For example, site-screening criteria can be used to identify various environmental criteria or project objectives (Box 6.3)

Box 6.3 Mackenzie Gas Project Alternatives Assessment

In 2003, Imperial Oil Resources Ventures Limited, Shell Canada, ConocoPhillips, ExxonMobil Canada, and the Mackenzie Valley Aboriginal Pipeline Limited Partnership submitted a preliminary information package describing the proposed Mackenzie Gas Project and its related environmental and socio-economic issues.

The proposed Mackenzie Gas Project is one of the largest energy development projects in North America. It involves the development of natural gas reserves on the Mackenzie River Delta in the Northwest Territories, natural gas processing, and the construction of a pipeline for natural gas shipment through the Mackenzie Valley to northern Alberta. At an estimated construction cost of $7 billion, the proposed pipeline will be approximately 1,220 kilometres long and will transport 34 million cubic metres of natural gas per day. More than 8,000 workers will be employed at peak construction, with an estimated 150 persons employed to operate the pipeline system. Up to 32 communities will be affected by the project, including 26 in the Northwest Territories and six in northwestern Alberta.

Given the project's magnitude and northern location, it was subject to EIA under three jurisdictions, the Canadian Environmental Assessment Act, the Mackenzie Valley Resource Management Act, and the Western Arctic Claims Settlement Act. A joint review panel was commissioned to administer the environmental impact review. Construction of the proposed project is expected to be completed and operational by 2009. All onshore and offshore production wells in the region will be connected to the pipeline by 2027.

Various alternatives were considered in the project EIS, including alternatives to the project (functionally different ways of producing, processing, and transporting gas to southern markets) and alternative means (technically and economically feasible ways to implement the proposed project). With regard to 'alternatives to', the proponent argued that there were no viable alternatives to the development and transportation of Mackenzie Delta gas to southern markets that were as advanced as the proposed project. The only true 'alternative to' was the no-go alternative, the result of which, according to the proponent,

continued

would be the loss of benefits and opportunities for both the project investors and northern residents.

With regard to 'alternative means', several options were considered, including alternative locations for facilities and infrastructure sites, route alternatives within the preferred pipeline corridor, and alternative methods for construction and reclamation. Of considerable concern, in terms of both environmental and community impacts, was the route selected for pipeline construction. On the basis of field survey, baseline data collection, and discussions with local communities, the preliminary pipeline route and alternative were divided into 36 segments for alternative route evaluation. Evaluation criteria for preferred route selection included route placement (reducing pipeline length), watercourse crossings (reducing the number, complexity, and width of crossings), geotechnical considerations (avoiding steep or unstable slopes), environmental impacts (land-use patterns, critical wildlife habitat), construction considerations (reducing length of steep slopes and grading required), community interests (community feedback), and cost (relative cost of route alternatives).

Alternative route assessment occurred over three phases: first, in August 2002 to scope preliminary route options based on terrain, physical limitations, and traditional land-use patterns; second, in September 2002 based on watercourse crossings to narrow and refine the preferred route options; and third, in August 2003 to confirm the preferred route mapping and to align the final route with proposed pipeline facilities and infrastructure.

Source: Imperial Oil Resources Limited 2004.

Candidate alternatives must be systematically compared to identify the preferred option among those that remain for detailed impact assessment. Several methods are available for evaluating alternatives (Table 6.1); the objective is to compare across alternatives using similar criteria or objectives. For example, alternatives can be evaluated using a simple rating or ranking procedure based on potential economic, social, or environmental impacts or contributions (Table 6.2). In cases where the selection among alternatives is not straightforward, assessment criteria may need to

Table 6.1 Selected Methods for Evaluating Alternatives*

Contingent ranking	Objectives hierarchies
Contingent valuation	Opportunity costs
Cost-benefit analysis	Paired comparisons
Decision trees	Performance measures
Expert systems	Queuing models
Life-cycle assessment	Social choice theory
Linear programming	Spatial decision support systems
Moment estimation methods	Utility functions
Multi-attribute analysis	Weighted scoring
Networks	Willingness to pay
Net present value	

*Many of these are equally applicable to impact prediction.

Table 6.2 Sample Alternative Site Evaluation Procedure

Candidate sites	*Minimize economic cost*	*Minimize impacts on water quality*	*Topographic suitability*	*Ease of access*	*Minimize community impacts*	*Minimize wildlife/ habitat impacts*	*Total score*
A	3	4	5	3	2	1	18
B	3	2	2	3	2	1	13
C	4	3	3	5	4	4	23
D	2	4	2	2	2	4	16

Legend: 5 = site meets criterion in full
4 = intermediate decision
3 = site meets criterion in part
2 = intermediate decision
1 = site does not meet criterion

be weighted to determine their relative importance (see Chapter 8). In short, the environmental and socio-economic effects of alternative means, including technical and economic feasibility, should be identified and clear justification provided for the preferred project alternative.

Valued Ecosystem Components

A large volume of baseline data could be collected during project scoping to characterize the biophysical (Table 6.3) and human environments (Table 6.4), and a long list of techniques for collecting that data could be employed from sources such as census data, historical records, land-use plans, field surveys, and sampling procedures. However, EIA scoping is not simply about undertaking data collection for the purpose of an 'environmental study'. Rather, scoping should establish clear boundaries for spatial and temporal assessment according to the important components of the environmental baseline. This includes limiting the amount of information to be gathered in baseline studies to a manageable level and identifying specific objectives and indicators to guide the assessment. In this way, establishing the assessment boundaries focuses baseline studies, making them more efficient for and relevant to subsequent stages of the assessment process (Figure 6.2).

Scoping valued ecosystem components. Comprehensive baseline studies can often be a waste of time and resources. Environmental baseline studies must be undertaken within the context of clearly defined scope and objectives; otherwise, too much information is acquired that is often of too little use. While it is important to ensure that all concerned or potentially affected environmental components are given consideration, attention should focus on the **valued ecosystem components (VECs)** most likely to be affected. VECs are aspects of the environment, physical and human, considered to be important from scientific or public perspectives and thus warranting detailed consideration in the impact assessment. In the Mackenzie Gas

Table 6.3 Scoping the Biophysical Environment

Air
- current pollutant concentration
- pollutant dispersion
- emission levels
- emission types
- temperatures
- windspeeds and directions

Soil
- erosion rates
- moisture content
- fertility
- organic matter
- electrical conductivity
- chemical composition
- stability
- soil pollutants

Terrestrial
- level of fragmentation
- widlife populations
- vegetation cover and composition
- air- and water-borne pollutants
- levels of light pollution
- vegetation health

Water
- surface quantity
- surface water withdrawal
- groundwater quantity
- groundwater withdrawal
- chemical content
- turbidity
- stream flow
- bank stability
- levels of eutrophication
- current pollution discharges
- fish and fish habitat

Coastal zone
- water temperature
- flood frequency
- tidal activity
- sedimentation
- marine resource populations
- bank or cliff stability

Table 6.4 Scoping the Human Environment

Economics
- local and non-local employment
- labour supply
- wage levels
- skill and education level
- retail expenditures
- material and service suppliers
- regional multiplier
- tourism

Demographics
- population
- population characteristics (family size, income, ethnicity)
- settlement patterns

Health
- quality of life (actual and perceived)
- medical standards
- worker death or injury rates
- current disease transmission
- mental and physical well-being

Housing
- public and private housing
- house prices
- homelessness and housing problems
- density and crowding

Local services
- educational services
- health services
- community services (police, fire)
- transportation services and infrastructure
- financial services

Socio-cultural
- family life
- seasonality of employment
- culture and belief systems
- crime rates, substance abuse, divorce rates
- community conflict and cohesion
- traditional foodstuffs
- community perception
- gender relations

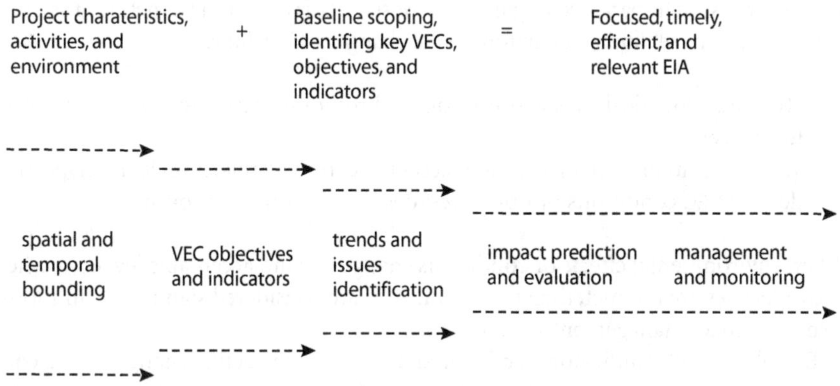

Figure 6.2 Scoping the environmental baseline for focused and informative EIA.

Project, for example, wildlife VECs in the natural gas pipeline corridor were identified on the basis of their regulatory status, ecological importance, socio-economic importance, and conservation concern. In the Voisey's Bay EIS, discussed in Chapter 1, 17 key VECs, including water, caribou, plant communities, Aboriginal land use, and family and community, were identified on the basis of social, economic, regulatory, and technical values and concerns. In addition, the EIS identified certain VECs that were also 'pathways' for environmental effects that might lead to effects or impacts on other VECs. For example, the VEC 'water' was identified as a biophysical effects pathway that could potentially affect other VECs: freshwater and fish habitat, waterfowl, and seabirds.

To ensure that baseline studies are purposeful and not simply a compilation of information about VECs, at least two questions should be asked about each VEC prior to moving forward and collecting detailed baseline environmental data:

1. Is the VEC likely to be affected, directly or indirectly, by project activities? If not, then there is no need to compile a comprehensive baseline about that VEC. This is not to say that the VEC should be disregarded, since it may still be of significant public concern and warrant some consideration in the impact assessment process.
2. Is it possible to predict impacts on the VEC or on indicators of that VEC or to relate either quantitatively or qualitatively project-induced change to VEC conditions? If not, the VEC should be given relatively low priority unless it is a VEC of significant public value.

Identify VEC objectives and indicators. Once VECs are selected, it is important to identify **VEC objectives** and **VEC indicators**. Such objectives and indicators not only help to focus baseline studies, but they are also important for impact prediction, monitoring, and impact management. First, establishing objectives or criteria for VECs is critical to evaluating the environmental impacts of the proposed project. These objectives or criteria can represent the specific parameters, guidelines, or standards that must be met, such as a carrying capacity or set limits of environmental change, and include:

- absolute ecological or socio-economic thresholds: carrying capacity, limit of tolerance;
- acceptable limits of change: what is acceptable from a broader societal perspective;
- desired VEC conditions or objectives: desired outcomes or conditions.

When environmental effects or conditions change beyond acceptable levels, or when VEC objectives are not met, then project impacts are considered significant, and some form of impact management is necessary.

Establishing VEC indicators is critical to understanding actual change in VEC conditions or, more important, to providing an early warning of potential impacts. VEC indicators are measurable parameters of change that are in some way related to the VEC of concern and in some cases may be VECs themselves. The EIS for the Jack Pine oil sands mine in northeastern Alberta referred to such indicators as 'KIRs', or 'key indicator resources'. Such indicators might include specific parameters of air quality, water quality, or employment rates. VEC indicators are the most basic tools for analyzing environmental change—they essentially allow decision-makers to gauge environmental change efficiently by avoiding impact 'noise' and focusing on parameters that are responsive to change, generate timely feedback, and can be traced effectively over space and time. Good VEC indicators are, at a minimum:

- measurable, either in a qualitative or quantitative fashion;
- indicative of the VEC of concern;
- sensitive and detectable in terms of project-induced stress;
- appropriate to the spatial scale of the VEC of concern;
- temporally reliable;
- diagnostic to change;
- standardized across development projects.

Establish Assessment Boundaries

Spatial bounding. What are the spatial limits of the assessment? In delineating the spatial limits of assessment, some consideration must be given to scale. João (2002) suggests that scale has two interrelated meanings relevant to EIA: first, scale as the *spatial extent* of the assessment; second, scale as the amount of *geographic detail*. When the spatial scale of an EIA covers a very large area, a high degree of geographic detail or granularity is less feasible. In other words, as the geographic boundary of an assessment increases, granularity typically decreases. This is an important consideration in spatial bounding, since different types of scales and levels of detail are often required to understand project interactions with different environmental receptors. Therefore, different environmental receptors should be examined at different geographic scales, and for any given receptor there are various scales at which the receptor can be assessed (Box 6.4). For example, air quality can be measured across a metropolitan area or an airshed or within the confines of a particular administrative space. In other words, factors affecting air quality can be assessed

Box 6.4 Principles for Spatial Bounding

- Boundaries must be large enough to include relationships between the proposed project, other existing projects and activities, and the affected environmental components (Cooper 2003).
- The scope of assessment should cross jurisdictions if necessary and allow for interconnections across systems (Shoemaker 1994).
- Natural boundaries should be respected (Beanlands and Duinker 1983).
- Different receptors will require assessment at different scales (Shoemaker 1994).
- Boundaries should be set at the point where effects become insignificant by establishing a maximum detectable zone of influence (Scace, Grifone, and Usher 2002).
- Both local and regional boundaries should be established (Canter 1999).

Geographic boundaries for any particular assessment will vary depending on a number of factors, including the nature of the project itself, sensitivity of the receiving environment, nature of the impacts, extent of transboundary impacts, availability of baseline data, jurisdictional boundaries and cooperation, and natural physical boundaries.

across a variety of **functional scales**, ranging from point source to local, to regional (including regions of various sizes), to national (Chagnon 1986).

It is important to distinguish between two types of information: the activity information and the receptor information (Irwin and Rodes 1992). **Activity information** characterizes the types of effects a project might generate, such as habitat fragmentation or increased use of non-renewable resources. **Receptor information** refers to the processes resulting from such effects. Both of these must be taken into consideration when delineating spatial boundaries. For example, the Canadian Environmental Assessment Agency's operational policy for spatial bounding regarding offshore oil and gas exploratory drilling suggests that the spatial area of an assessment should be the 'composite' of the spatial area of all affected environmental receptors (Figure 6.3)

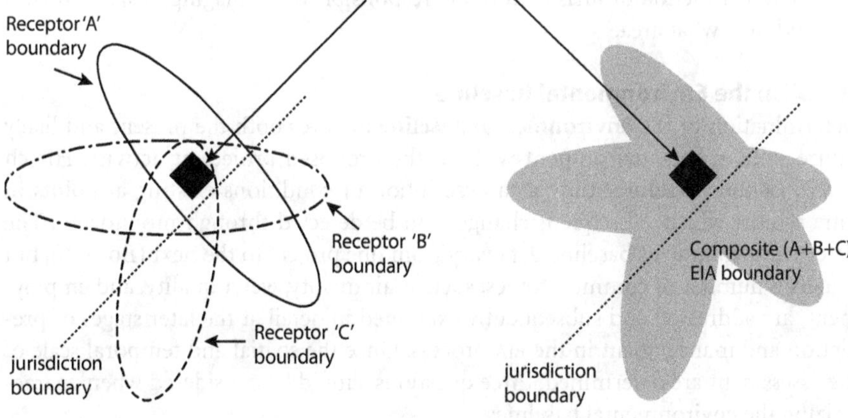

Figure 6.3 Spatial bounding based on composite environmental receptor information

and should consider uncertainties concerning the precise location of proposed activities and effects, as well as the need to reconsider individual project bounding when multiple projects are being conducted in adjacent resource areas. Several of the methods identified in Chapter 4 can be used to assist in spatial bounding, including Geographic Information Systems, network and system analysis, matrices, public consultations, quantitative and physical modelling, and expert opinion.

Temporal bounding. What are the temporal limits of the assessment? The temporal scale should include the past, present, and future. This involves consideration of previous and current activities in the project region affecting environmental conditions as well as potentially foreseeable activities. For example, a proposed hazardous landfill site might involve consideration of previous industrial activities that have affected the baseline environment as well as current and proposed land-use activities. How far into the past and how far into the future will depend on the quality and quantity of data available, the certainty associated with project environmental conditions, and what the assessment is trying to achieve. Examining past conditions may be as simple as examining land-use maps or census data, and it may be feasible to incorporate 50 years of historical data if deemed necessary for the proposed project. Establishing a future boundary for the assessment may be based on the end of the operational life of a project, after project decommissioning and reclamation, or after affected environments recover. Future boundaries are much less certain, since data are often hypothetical or based on a range of future scenarios (Box 6.5). Typically, future boundaries do not exceed five years beyond project decommissioning.

Jurisdictional bounding. What are the jurisdictional limits of the assessment? Certain impacts may spread across administrative boundaries and therefore affect environments within different jurisdictions. In other cases, EIA screening may suggest that the nature of the proposed project involves both regional and national authorities. There are no general rules for establishing jurisdictional boundaries; however, it is important to consider:

- what jurisdictions are affected by the projects;
- what jurisdiction or jurisdictions have decision-making authority;
- what jurisdiction or jurisdictions are responsible for managing project impacts and over what area.

Establish the Environmental Baseline

Determination of the **environmental baseline** involves both the present and likely future state of the environment without the proposed project or activity. Hirsch (1980) defines **baseline study** as a description of conditions existing at points in time against which subsequent changes can be detected through monitoring. The nature and volume of baseline data vary from one project to the next (Box 6.6), but usually a number of common topics, such as air quality, water quality, and employment, are addressed and subsequently examined in detail at the later stages of prediction and management in the EIA process. Once the spatial and temporal scale of the assessment are determined, three questions should be considered when characterizing the environmental baseline:

Box 6.5 Temporal Bounding in the Eagle Terrace Subdivision Assessment

Canmore, Alberta, is located in the Bow River Valley east of Banff National Park—one of Canada's most popular tourist destinations. Associated with the continued growth in tourism, however, is an increased demand for residential and visitor facilities, resulting in increased pressure for housing development. In 1996, a proposal was submitted to the town of Canmore for the development of the 67-hectare Eagle Terrace residential subdivision on the slopes of the mountain valley. The proposed project would join the existing subdivisions already located near the town.

Given the involvement of a national park, the impact assessment was carried out pursuant to federal requirements. Spatial boundaries for the Eagle Terrace assessment were based on the availability of a vegetation map that covered enough of the mountain valley to include a sufficient number of actions associated with the project under review, existing land-use activities, and the information necessary to assess potential effects on wildlife. The temporal boundaries of the assessment, however, were much less certain and based on a number of potential future scenarios, including:

- pristine scenario—a future state characteristic of conditions prior to any or extensive human development, which was simulated by removing the footprint of all developments from a GIS database;
- current scenario—assumption of the existing development, land-use, and impact conditions;
- future 'without action' scenario—assumption of future conditions predicted to occur in the absence of the proposed development;
- future 'with action' scenario—assumption of future conditions predicted to occur with the proposed action under review.

The project EIS concluded that the proposed development (future 'with action' scenario) would incrementally contribute to the current losses of habitat but less than 2 per cent of what had already occurred.

Source: Hegmann et al. 1999.

- What do we need to know about the baseline environment to make a decision on the project?
- What are the relevant background conditions that have influenced the current environment?
- What is the likely baseline condition in the future in absence of the project?

A key component then of establishing the environmental baseline is identifying and establishing baseline trends.

Delineating VEC conditions and baseline trends. Once VECs are defined and their objectives and indicators determined, the current conditions of those VECs and, ideally, trends over space and time should be determined. This phase of the baseline is also referred to as the retrospective phase of impact assessment. Current VEC conditions, or the current state of VEC indicators, are determined. This may involve, for

Box 6.6 Experiences with Collecting Baseline Data

Collecting baseline data may require several seasons or years to quantify sufficient ranges of natural variation and directions and rates of change, yet these data are often critical to effective impact identification and management. In the case of the La Grande-2A and La Grande-1 hydroelectric powerhouses located on the La Grande River in Quebec, a three-year program was initiated to establish baseline environmental conditions between 1987 and 1990. However, this is perhaps an exception to conventional practice in that baseline studies are rarely done sufficiently. Time constraints in EIA usually preclude lengthy survey and data collection programs, and impact predictions typically have to rely on existing data. In frontier areas, even existing data can be limited. In the case of the Ekati Diamond Project in the Northwest Territories, for example, most biophysical impact predictions and associated mitigation measures in the EIS were based on data collected during just one field season.

Sources: Based on Mulvihill and Baker 2001; Denis 2000.

example, determining current methyl mercury concentrations in the river system at the site of a proposed large-scale hydroelectric development project as an indicator of water quality or determining current physician-to-patient ratios in the affected community as an indicator of community health and well-being.

The objective is to determine how such indicators looked in the past, how they changed over time, and what types of stressors resulted in or triggered such changes. Where spatial and temporal trends can be identified, this information can be used in the prospective, or impact prediction, stage of EIA and also to gain a sense of what the project environment or VECs might look like in the future, assuming past trends prevail, in the absence of the proposed project.

In cases where VECs or VEC indicators are not amenable to interpolating trends, information on past conditions can be used alongside current conditions as benchmarks against which potential project-induced stress can be identified and assessed. If, for example, the trend in the project environment has been towards decreasing physician-to-patient ratios and the health care system is currently at or near capacity, then any induced stress caused by an increase in health care demand will be significant in terms of community health and well-being and thus warrant detailed consideration in impact assessment and mitigation.

Identify Potential Impacts and Issues

Once the project is described and the baseline established, potential project impacts and areas of concern can be identified. The objective is to highlight for further and more intensive study those issues and impacts that will form an important part of the decision as to whether a project will be approved and under what conditions. Methods for impact identification were discussed in Chapter 4. As suggested earlier, perhaps the most common method for presenting and organizing potential impacts is an impact matrix, which identifies project activities and interactions with environmental components or VECs and characterizes the nature of those interactions.

The objective is to identify key areas of concern for further impact prediction, assessment, evaluation, and monitoring. Several of the methods identified in Chapter 4 can be used to assist in impact identification, including Geographic Information Systems, network and system analysis, matrices, public consultations, quantitative and physical modelling, and expert opinion. Perhaps the most common methods, however, involve a combination of expert judgment, public consultations, and scoping checklists or simple scoping matrices (Figure 6.4).

Affected Components

Actions	wildlife	agricultural land	air	water	noise
Construction Phase					
dredging	U I 1(–) S L	U ? 1(–) S L	U ? 1(–) ? ?	C I 3(–) L R	P I 1(–) S L
site clearing	P I 3(–) P L	C I 3(–) S R	U D 1(–) L R	P D 1(–) S R	P I 1(–) S L
lake dewatering	P I 2(–) P L	U D 1(–) S L	U ? ?(?) ? ?	C I 5(–) P L	U ? ?(?) S L
road construction	C I 3(–) L R	U ? 1(+) L R	P I 1(–) S R	P D 2(–) S L	C I 1(–) S L
traffic and equipment	P I 2(–) L R	U D 1(–) S R	P I 1(–) L R	P D 1(–) S L	C I 2(–) L R

Legend:

likelihood	time frame
magnitude (direction)	
duration	extent

Likelihood: C = certain, P = probable, U = unlikely, ? = unknown
Time: I = immediate, D = delayed, ? = unknown
Magnitude: 5 = major, 3 = moderate, 1 = minor
Direction: + = positive, – = negative, ? = unknown
Duration: P = permanent, L = long-term, S = short-term
Extent: L = local, R = regional, N = national

Figure 6.4 Sample project scoping matrix.

SCOPING CONTINUATION

A good scoping process is said to have a positive ripple effect throughout the rest of the EIA process. Scoping ensures that all of the necessary factors are considered in the assessment and at the same time sufficiently narrows the scope of the assessment to focus only on centrally important elements. While scoping is typically identified as a single phase of the EIA process, it is important to note that scoping is an ongoing activity in EIA and should continue in the post-project implementation and monitoring phases. Good-practice scoping is a process by which the environment is continuously scanned to detect signs of adverse environmental changes or variables of importance that may not have been considered during initial project scoping exercises.

KEY TERMS

activity information
alternative means
alternatives to
baseline study
closed scoping
environmental baseline
functional scale
open scoping
receptor information
scoping
valued ecosystem components (VECs)
VEC indicators
VEC objectives

STUDY QUESTIONS AND EXERCISES

1. What is the purpose of EIA scoping?
2. What are valued ecosystem components?
3. Obtain copies of topographic and political maps of your region. Suppose a new coal-fired electricity generating plant is being proposed to meet a growing demand for electricity:
   a) Identify a project location on the map sheet.
   b) Which VECs would you identify in the region as important to consider? Provide a statement as to why these environmental components are classified as VECs. In other words, what makes them so important that you would include them in the assessment?
   c) Given current VEC conditions, what objectives would you attach to each VEC?
   d) Identify a list of indicators that you might use to assess and monitor the conditions of each VEC.
   e) What spatial boundary or boundaries would you establish for the assessment? Sketch these boundaries on a copy of the map. Identify the criteria you used to determine these boundaries.
   f) What temporal boundaries might you consider?
   g) Brainstorm a number of feasible alternatives that might be considered in the project assessment, including alternative project locations.
4. Obtain a completed project EIS from your local library or government registry, or access one on-line. Identify the number and types of alternatives, if any, considered in the assessment. Were these alternatives assessed and if so, to what extent? What VECs were identified? Compare your results with the findings of others.

REFERENCES

Beanlands, G.E., and P.N. Duinker. 1983. 'Lessons from a decade of offshore environmental impact assessment'. *Ocean Management* 9 (3/4): 157–75.

Canter, L. 1999. 'Cumulative effects assessment'. In J. Petts, Ed., *Handbook of Environmental Impact Assessment*, v. 1, *Environmental Impact Assessment: Process, Methods and Potential.* London: Blackwell Science.

Chagnon, S. 1986. 'Atmospheric systems: Management perspective'. In Canadian Environmental Assessment Research Council (CEARC), Ed., *Cumulative Environmental Effects Assessment in Canada: From Concept to Practice.* Calgary: Alberta Society of Professional Biologists.

Cooper, L. 2003. *Draft Guidance on Cumulative Effects Assessment of Plans.* EPMG Occasional Paper 03/LMC/CEA. London: Imperial College.

Denis, R. 2000. 'Lessons derived from the environmental follow-up programs on the La Grande Rivière, downstream from the La Grande-2AQ generating station, James Bay, Quebec, Canada'. Paper presented at the International Association for Impact Assessment annual meeting, June, Hong Kong.

Hegmann, G., et al. 1999. *Cumulative Effects Assessment Practitioners Guide.* Prepared by AXYS Environmental Consulting and CEA Working Group for the Canadian Environmental Assessment Agency, Hull, QC.

Hirsch, A. 1980. 'The baseline study as a tool in environmental impact assessment'. In *Proceedings of the Symposium on Biological Evaluation of Environmental Impacts.* Washington: Council on Environmental Quality and Fish and Wildlife Service, Department of the Interior.

Imperial Oil Resources Limited. 2004. *Environmental Impact Statement for Mackenzie Gas Project.* Ottawa: National Energy Board and Joint Review Panel.

Irwin, F., and B. Rodes. 1992. *Making Decisions on Cumulative Environmental Impacts: A Conceptual Framework.* Washington: World Wildlife Fund.

João, E. 2002. 'How scale affects environmental impact assessment'. *Environmental Impact Assessment Review* 22: 289–310.

Mulvihill, P., and D. Baker. 2001. 'Ambitious and restrictive scoping: Case studies from northern Canada'. *Environmental Impact Assessment Review* 21: 363–84.

New Zealand Ministry for the Environment. 1992. *Guide for Scoping and Public Review Methods in Environmental Impact Assessment.* Wellington: Ministry for the Environment.

Noble, B.F. 2004. 'Integrating strategic environmental assessment with industry planning: A case study of the Pasquia-Porcupine forest management plan, Saskatchewan, Canada'. *Environmental Management* 33 (3): 401–11.

Scace, R., E. Grifone, and R. Usher. 2002. *Ecotourism in Canada.* Ottawa: Canadian Environmental Advisory Council.

Shoemaker, D. 1994. *Cumulative Environmental Assessment.* Waterloo, ON: University of Waterloo, Department of Geography.

SMLP (Saskfor-MacMillan Limited Partnership). 2004. *Twenty-Year Forest Management Plan and Environmental Impact Statement for the Pasquia-Porcupine Forest Management Area.* Assessment main document. Regina: SMLP.

CHAPTER 7

Predicting Environmental Impacts

IMPACT PREDICTION

Impact assessment requires both forethought and foresight about the potential implications of a proposed development project. Consequently, impact prediction is fundamental to EIA. A prediction is a statement specifying, without direct measurement, the past, present, or future condition of a particular VEC or VEC indicator, given particular system characteristics (Duinker and Baskerville 1986). Not all aspects of project impacts should be predicted, since this would be neither feasible, given time and resources, nor desirable, given the volume of information that can be effectively processed in any single EIA. Prediction involves identifying potential changes in *impact indicators* of environmental receptors (VECs) identified during EIA scoping and requires:

- determining the initial or reference state (baseline) and baseline trends;
- predicting the future state in the absence of project development (the future baseline);
- predicting the future state with project development.

Understanding baseline trends is important to understanding the true nature of the predicted impact. For example, if employment in the area for which the project is proposed has been steadily increasing over the previous five years, then this baseline trend should be predicted in the absence of the project in order to allow for a more informed decision concerning the significance of the predicted employment contributions of the proposed project. Similarly, if a project is predicted to have a negative effect on fish populations, then this information should be considered in light of recent and predicted trends in fish populations in the area—factoring out, where possible, impacts from other sources and changes due to natural variability.

Predicting the potential environmental impacts of proposed projects is a complex and highly uncertain task, since many cause–effect relationships are unknown and physical and human environments are 'moving targets'. Thus, Morris and Therivel (2001) suggest that impact prediction requires several elements, including:

- sound understanding of the nature of the proposed undertaking;
- knowledge of the outcomes of similar projects;
- knowledge of past, present, or approved projects whose impacts may interact with the proposed undertaking;

- predictions of the project's impacts on other environmental and socio-economic components that may interact with those directly affected by the project;
- information about environmental and socio-economic receptors and how they might respond to change.

BASIC REQUIREMENTS OF IMPACT PREDICTIONS

Impact prediction should provide insight into the specific characteristics of potential impacts. These characteristics include:

- the nature of the predicted impact (for example, adverse, additive, antagonistic);
- temporal characteristics;
- magnitude, direction, and spatial extent;
- degree of reversibility;
- the likelihood that the predicted impact will actually occur.

These are basic principles of impact classification and are critical to determining impact significance, as discussed in the next chapter. In addition, two underlying principles essentially determine the usefulness of impact predictions for managing project impacts: accuracy and precision.

Accuracy and Precision

Impact predictions such as 'slight reduction' or 'minor effect' are of little value for understanding, monitoring, and managing the actual effects of project development. It is possible to generate quite accurate predictions when such predictions are couched in vague language. The value of such predictions, however, is limited in terms of avoiding or managing project effects. **Accuracy** refers to the extent of system-wide bias in impact predictions, or the closeness of a predicted value to its 'true' value. **Precision** refers to the level of preciseness or exactness associated with an impact prediction (Figure 7.1). Accuracy, then, depends on the level of precision required. One can be consistently accurate when predictions are imprecise. As the desired level of precision increases, then the possibility of inaccuracy in impact prediction also increases. The fact that an impact prediction is accurate does not necessarily constitute a useful impact prediction if the level of precision is relaxed. However, if more precise impact predictions are desired, then accuracy may be forfeited (De Jongh 1988).

Although prediction is the cornerstone of EIA, little attention is given to the manner in which predictions about anticipated effects are phrased. There are two dimensions to predictive accuracy: (1) the logical validity of the impact prediction; (2) the relative severity of the actual versus predicted impact. Buckley (1991) explains that in cases where actual impacts prove less severe than anticipated, predictions may still be logically incorrect if they were expressed as 'impact parameter will be equal to predicted numerical value' rather than 'impact parameter will be less than or equal to predicted numerical value.' However, whether an impact prediction is logically correct is perhaps less meaningful than the accuracy of the relative severity of actual versus predicted impacts—that is, impacts less or more severe than initially predicted.

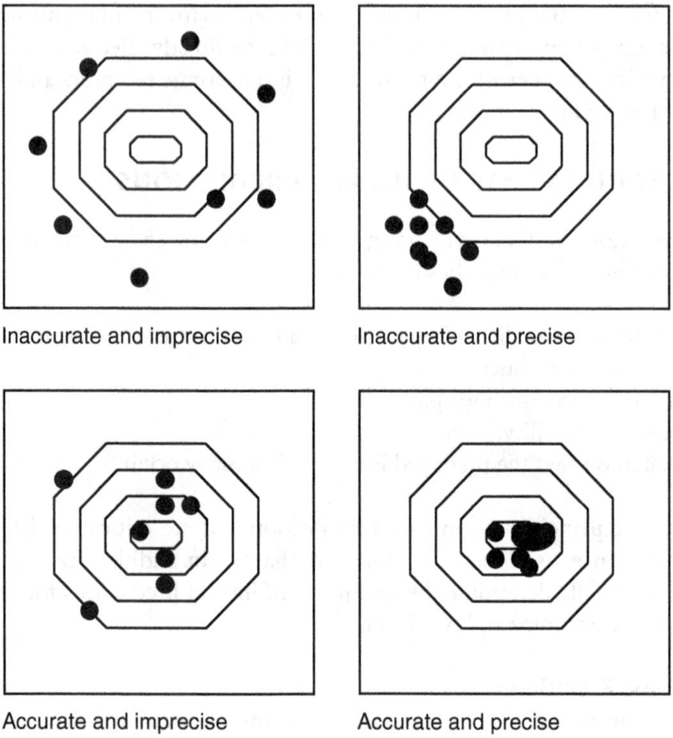

Figure 7.1 Accuracy versus precision.

TOOLS FOR IMPACT PREDICTION

There is some confusion in many EIA texts and guides concerning the role of checklists and matrices. While checklists and matrices are used to assist impact predictions as a means of organizing data and information, they are *not* tools for predicting impacts; rather, they are methods for presenting and organizing predictions. The choice of predictive tool depends on the nature of the impact indicator in question. For example, when predicting the *magnitude* or severity of an impact, the approach may be qualitative and based on expert judgment. When predicting the concentration of a particular pollutant, more quantitative-based modelling tools might be used. Tools used for impact prediction cover a wide spectrum of economic, social, ecological, and physical factors. Morris and Therivel (2001) provide a detailed treatment of predictive tools. While there is no single best approach to classifying tools used for prediction in EIA, there are essentially seven broad classifications:

- models
- extrapolation
- experimental
- analogue
- judgmental

- scenarios
- spatial

  (see Sadar 1996; Dale and English 1999; Glasson, Therivel, and Chadwick 1999; Harrop and Nixon 1999; Greig, Pawley, and Duinker 2002; Duinker and Greig 2007).

This classification certainly does not cover all of the methods and techniques available for impact prediction in EIA but represents a cross-section of those most commonly used.

Modelling Tools

The use of modelling is common practice in EIA for predicting the future state of environmental and economic variables when sufficient data are available to support such tools. Modelling creates simplified representations of the system under investigation, including causal mechanisms, and is based on explicit assumptions about the behaviour of the system. A significant challenge to modelling approaches, however, is that they are typically very 'data demanding'. The required input to support reliable modelling approaches is not always readily available from environmental baseline studies, and data may require several field seasons to collect.

Mechanistic models describe cause–effect relationships in the project environment using flow diagrams or mathematical equations. Examples include Gaussian dispersion models for predicting rates and extents of pollution fallout, traffic noise generation models, population growth models, and economic impact models. Quantitative models used in EIA can be divided into two broad categories—deterministic and stochastic. **Deterministic models** depend on a fixed relationship between environmental components and provide a single solution for the state variables. For example, the **gravity model** of spatial interaction, used to predict population flows, depends on a fixed and inverse relationship between mass (population size) and distance. The system under consideration is thus at any time entirely defined by the initial conditions and model inputs chosen—uncertainty and variability are not explicitly modelled. With a given starting point, therefore, the outcome of the model's response is necessarily the same and is determined by the mathematical relationships incorporated in the model. **Stochastic models** are probabilistic—that is, they give an indication of the *probability* of an event or events within specified spatial and temporal scales and take into consideration the presence of randomness in one or more variables. In other words, a stochastic model includes variability in the model parameters as a function on spatial and temporal aggregation, random variability, and changing environmental conditions. Thus, rather than giving a single-point estimate, stochastic models provide a probability distribution of possible estimates or outcomes—such as the likelihood of a flood as a result of river control systems.

Mass **balance models**, or mass balance equations, are another type of modelling tool used in EIA. Balance models identify inputs and outputs for specified environmental components, such as soil moisture volume or surface runoff, and are essentially comprised of inputs, storage, and outputs for a particular environmental system (Box 7.1). For example, inputs to a particular system might include water and energy, whereas outputs include waste water and outflow. Balance models, such as surface

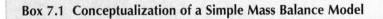

Box 7.1 Conceptualization of a Simple Mass Balance Model

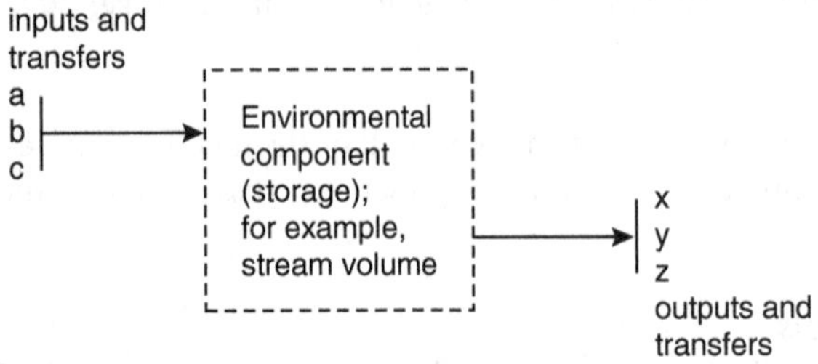

When the system is 'balanced', a + b + c = x + y + z. Changes in the environmental component can be predicted by examining potential project impacts on inputs and transfers, which will be evidenced by changes in subsequent outputs and transfers.

hydrological models, are particularly useful for predicting physical changes in environmental phenomena when predicted changes equal the sum of the total inputs minus the sum of the total outputs (Glasson, Therivel, and Chadwick 1999). A well-known example is a site water budget, which predicts the impacts of a proposed development on the input, storage, and output of water available in a particular sub-basin.

Extrapolation Tools

Extrapolation, or trends analysis, uses assumptions about fundamental causes underlying an observed phenomenon in order to project beyond the range of available data and information. Extrapolation in EIA is typically based on the use of **statistical models**, simple linear or multivariate regression, and fitting data to a mathematical function. The primary use of statistical models is to test relationships between variables and to extrapolate data, not to determine cause–effect relationships. For example, based on hypothesized or proven statistical associations between neighbourhood housing density and quality of life, statistical models can be used to extrapolate changes in the quality of life resulting from a proposed large-scale residential housing development program.

Statistical models are also used to determine whether a statistically significant difference exists between the predicted changes in impact indicators due to project influence as opposed to natural changes that would occur in the absence of project development. A word of caution is in order, however, because a measure of statistical significance, such as a 95 per cent confidence interval, does not necessarily translate directly to human perception or acceptance. For example, depending on the nature of the predicted impact, the possibility of a 5 per cent error may be deemed significant enough not to allow the action to proceed—particularly if human health and safety are at risk. A particular class of statistical models of significant value to the modelling of dynamic environments is the multivariate model, which considers changes and interactions between multiple variables within a given environmental system.

Experimental Tools

Experimental tools allow data or experimentation in a field or lab setting to be applied to the project environment—for example, testing the impacts of a chemical agent on a water body or predicting emissions patterns based on laboratory-controlled wind tunnels. While useful, it is important to note that not all experimental lab and field conditions transfer well to the actual project environment, particularly with regard to human conditions and impacts. Experimental tools often adopt empirical models or models for which the structure is determined by the observed relationship among experimental data, as opposed to mechanistically relevant relationships.

Analogue Tools

Analogue techniques make impact predictions of a proposed development based on comparisons with the impacts of existing developments where conditions might be similar. Data might be collected and comparisons made on the basis of literature reviews, site visits, or expert opinion. A main drawback of analogous techniques, similar to ad hoc approaches, is that in many cases, experiences are not directly transferable from one project environment or one socio-economic setting to the next (Box 7.2).

Box 7.2 Analogue Techniques and the Hibernia Offshore Oil Platform Construction Project

In 1979, the Hibernia offshore oil field was discovered on the Grand Banks of Newfoundland. A submitted proposal to develop the oil field was subject to a panel review EIA under the then Canadian Federal Environmental Assessment Review process. Approval of the project was granted in 1986, and construction of a fixed gravity-base rig structure began at a dry dock in Bull Arm, Newfoundland, in 1990. The oil field is the fifth largest discovered in Canada, producing 7.6 million barrels of oil in 2004. Several companies have a share in the Hibernia megaproject: ExxonMobil Canada (33.125 per cent), Chevron Canada Resources (26.875 per cent), Petro-Canada (20 per cent), Canada Hibernia Holding Corporation (8.5 per cent), Murphy Oil (6.5 per cent), and Norsk Hydro (5 per cent).

Standing at half the height of New York's Empire State Building, the Hibernia oil-drilling platform was the first of its kind to be developed in Newfoundland and at the time was considered one of the most significant construction projects in North America. In the absence of direct experience with such a project in Newfoundland or on Canada's east coast, the main sources of information for EIA forecasting regarding the impact of the Hibernia project on population growth, employment, and infrastructure needs in St John's, Newfoundland, were other locales that had experienced sudden growth resulting from oil exploration—namely, in Aberdeen, Scotland, and Calgary, Alberta. Storey (1995) reports that comparisons of St John's with Aberdeen and Calgary led to the assumption that St John's would experience rapid population growth as a result of the Hibernia project. While such projections may have been correct for Aberdeen and Calgary, they failed to account for variations in the baseline social and institutional environment. For example, immediately following the first major oil discovery near Aberdeen in 1969, the population increased by

continued

2.5 per cent during the first half of the 1970s. For Aberdeen, this increase was considered rapid and problematic in comparison to previous growth rates, and local institutional and infrastructural capacities were not equipped to deal with such growth. However, in St John's during a similar time period, the population increased by more than 8 per cent, independent of oil development, and there were no major strains on institutions and infrastructure. Immediately following oil discovery, Calgary also experienced major population growth; however, it was primarily due to local labour shortages and the labour-intensive nature of the onshore oil industry. In contrast, St John's had a high unemployment rate and was involved in relatively capital-intensive offshore oil operations. Predictions of population growth and housing demand were thus overestimated.

When making predictions based on a comparison with similar projects, careful consideration must be given not only to the projects themselves but also to the nature of the biophysical and human environment in each project area.

Sources: Storey 1995; Storey and Jones 2003.

Judgmental Tools

In practice, judgmental tools are perhaps the most commonly used predictive tools in EIA. Judgmental tools can be classified as expert-based or non-expert-based, examples of which include **intention surveys** and the **Delphi technique**. Intention surveys are often used to survey individuals about what actions they plan to take in a given situation or what they might expect to do given particular circumstances. While this approach attempts to survey and incorporate the decisions of as many people as possible, it is a costly and time-consuming exercise. In addition, people may have difficulty in conceptualizing how a change might affect their actions or decisions.

The term 'Delphi technique' was coined by Kaplan, a philosopher working with the Rand Corporation who in 1948 headed a research effort directed at improving the use of expert predictions in policy-making (Woudenberg 1991). The name itself refers to the ancient Greek oracle at Delphi, where people seeking advice were offered visions of the future. The Delphi technique, developed in its contemporary form during the 1950s and early 1960s, is an iterative survey-type questionnaire that solicits the advice of a group of experts, provides feedback to all participants on the statistical summaries of the responses, and gives each expert an opportunity to revise his or her judgment (Figure 7.2). The Delphi typically undergoes a number of distinct phases or survey rounds. In the first survey round, an 'open-ended' questionnaire is sent to the panelists, who can then contribute additional information that they may feel is pertinent to the issue. In some cases, when the Delphi facilitator is seeking the opinion or the expert judgment of each individual regarding the issue(s) under consideration, a more structured questionnaire is used rather than an open-ended one. Responses are collected and summarized statistically and often incorporated into a number of analytical-based matrices. With this information at hand, the Delphi facilitator then compiles the round two questionnaires. The second-round and subsequent questionnaires typically include a reiteration of questions from the first questionnaire that engendered considerable discrepancies (e.g. conflict/non-consensus) and new questions based on issues and options raised by respondents during the first round. The process is repeated as many times as necessary until a desired level of consensus is reached. The typical Delphi procedure requires just three survey rounds, since consensus increases strongly over the first two survey rounds.

The Delphi technique is designed for use in situations in which the problem does not lend itself to precise analytical techniques or for which large data requirements and the lack of available quantitative data prohibit the application of traditional analytical

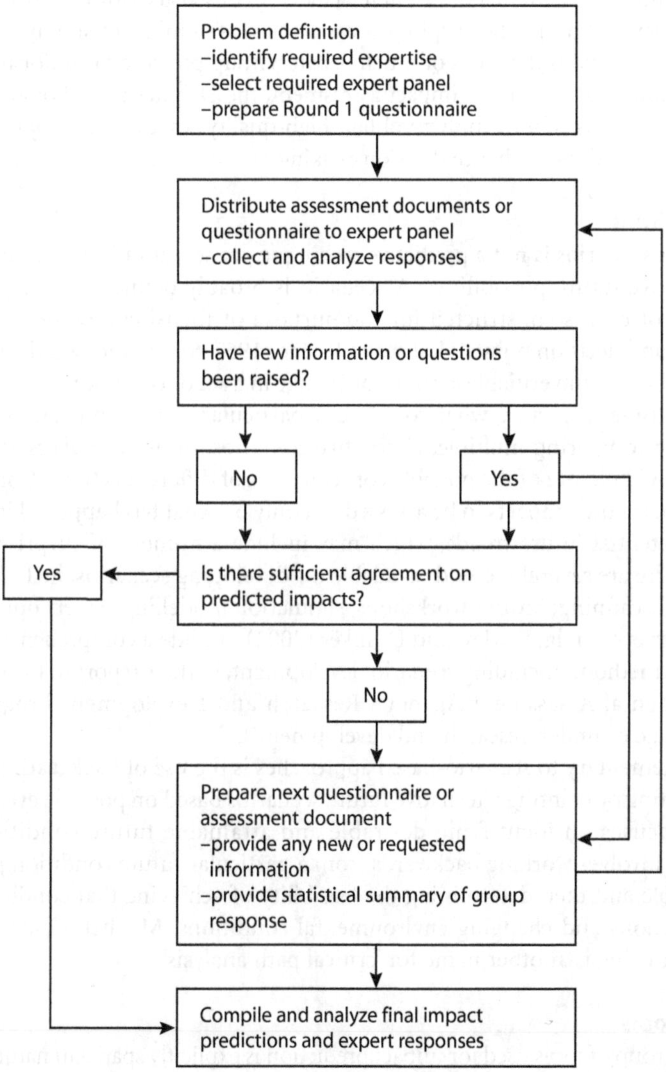

Figure 7.2 The Delphi process.

approaches but that can benefit from the subjective judgments of experts on a collective basis (Dalkey 1969). The technique has three key features: anonymity, iteration, and feedback, all of which are key to effective group decision-making. First, the Delphi technique allows the exchange of opinion among the experts while avoiding the pitfalls of face-to-face interaction, particularly conflict and individual dominance. Second, multiple iterations allow the 'least informed' participants to change their minds after hearing the group's response. Finally, feedback gives the experts who find the composite group judgments or the judgments of deviating experts more compelling than their own an opportunity to modify their decisions. For these reasons, the Delphi technique is extremely efficient in achieving consensus. However, whether consensus achieved through the Delphi process is 'oneness of mind' or simply the illusion of oneness of mind is the focus of much debate. Group pressure to conform does not reflect genuine agreement. In impact assessment, the Delphi's goal should not be to arrive at a consensus but simply to obtain high-quality responses and opinions on a given issue in order to enhance decision-making.

Scenario Tools

The use of scenarios is not a predictive technique per se but rather a way of looking at alternative future possibilities. A scenario is broadly defined as 'a hypothetical sequence of events constructed for the purpose of focusing attention on causal processes and decision points' (Kahn and Wiener 1967, 6). In other words, a scenario is a plausible but unverifiable account of change in a set of conditions over a defined period of time, depicting what could be if particular trends and rates of change unfold. By comparing multiple, alternative scenarios, decision-makers are able to obtain a vivid picture of the possible consequences of different actions. Comparison of scenarios allows impacts to be assessed not only on what has happened in the past but on potential future trends, which may include a number of surprises (Noble 2008). There are several methods available for developing scenarios, including environmental scanning, futures workshops, simulation modelling, expert opinion, and Delphi surveys. Greig, Pawley, and Duinker (2002) provide a comprehensive review of futures methods, including scenario development, in their report to the Canadian Environmental Assessment Agency's Research and Development Program (see www.ceaa.gc.ca under 'research and development').

Complementary to scenario-based approaches is the use of **backcasting**. Instead of forecasting or painting alternative future scenarios based on possible events, backcasting focuses on identifying desirable and attainable future conditions. The approach involves working backwards from a particular future condition judged to be desirable and then determining the feasibility of achieving that condition given project actions and changing environmental conditions. Mitchell (2002) explains that backcasting is another name for 'critical path analysis'.

Spatial Tools

Another group of tools used for impact prediction is explicitly spatial in nature. Spatial tools used in EIA for impact prediction often draw heavily on Geographic Information Systems (see Chapter 4) and range from simple overlay approaches whereby potential land-use conflicts can be 'predicted' to more complex spatial modelling and simulation exercises in which the distribution of impacts or phenomena are modelled over space—such as Marxan (see Box 7.3). Spatial tools for impact prediction have the advantage of presenting a 'visual' depiction of potential project effects and are particularly useful in supporting determinations of impact significance based on spatial extent. Similar to modelling-based approaches in general, however, spatial tools designed to simulate landscape patterns or change require significant data input.

PREDICTING IMPACTS ON THE BIOPHYSICAL ENVIRONMENT

Predicting the effects or impacts of project development on the biophysical environment has been at the heart of EIA since its formal inception under the US NEPA. In fact, as discussed in Chapter 1, most EIAs during the 1970s and well into the 1980s focused almost exclusively on predicting the biophysical impacts of project development within the local project area. In current practice, biophysical impact prediction continues to play the dominant role in EIA, but its focus has broadened considerably beyond the direct and immediate project environment to include secondary and ecosystem-based biophysical change.

What to Predict

Project developments often trigger four broad areas of change that affect biophysical systems:

- biological change
- chemical change
- physical change
- ecosystem change

Based on these changes, a number of specific effects and impacts often emerge (Table 7.1). There is no single, comprehensive list of biophysical impacts that should be considered in all EIAs, since the specific impacts to be predicted depend largely on the nature of the project and the sensitivity of the local environmental setting and are defined by the scoping process. Box 7.3 gives an example of an impact prediction using spatial tools for extrapolation and scenario analysis for biodiversity assessment in the Great Sand Hills, Saskatchewan.

Techniques

It is beyond the scope of this book to provide a comprehensive discussion of the full range of techniques available for predicting biophysical environmental effects. Numerous techniques exist for predicting such impacts, including plume models, water site budgets, wildlife population modelling, simulation modelling, dose-response factors, hydrographic extrapolation, and expert judgment (Table 7.2). Boxes 7.4 and 7.5 provide examples of two different techniques for predicting biophysical environmental impacts. Morris and Therivel (2001) give techniques for predicting biophysical environmental impacts much more comprehensive treatment.

Table 7.1 Selected Biophysical Environmental Impacts: Examples of What to Predict

Air quality impacts
- pollutant concentration (various)
- pollutant dispersion
- emission levels
- emission types
- temperature
- changes in wind speed

Soil quality impacts
- erosion
- moisture
- fertility (nutrient change)
- changes in organic matter
- electrical conductivity
- chemical change
- stability
- soil pollution

Terrestrial impacts
- fragmentation
- wildlife populations
- vegetation cover and composition
- air- and water-borne pollutants
- light pollution
- disturbance
- vegetative trampling

Water impacts
- surface quantity
- surface water withdrawal
- groundwater quantity
- groundwater withdrawal
- chemical change
- turbidity
- streamflow change
- bank erosion (flood risk)
- biological change (eutrophication, algal blooms)
- pollution discharge rates (assimilative capacity)
- biological resources (fish and fish habitat)

Coastal zone impacts
- water temperature
- flooding
- alterations to tidal activity
- sedimentation
- marine resource populations
- bank or cliff stability

Note that these impacts are based on scoping issues identified in Table 6.3 to illustrate the link between scoping baseline studies and focusing impact prediction.

Box 7.3 Predicting Impacts on Biodiversity in the Great Sand Hills, Saskatchewan

The Great Sand Hills of Saskatchewan comprises approximately 1,942 square kilometres of native prairie overlaying a more or less continuous surface deposit of unconsolidated sands, with five dune complexes that total 1,500 square kilometres. Also characterized by open grasslands and patches of trees and shrubs, the region is home to several game species and endangered, threatened, and sensitive species. Natural gas has been exploited in the region since the early 1950s and intensively since the 1980s, with approximately 1,500 gas wells now in the area. Livestock grazing has exerted a much longer-term and widespread influence on the landscape, the result of which is a network of permanent trails and vegetative trampling and erosion. In 2004, following decades of growing concern over the effects of natural gas development and ranching activities on the biological diversity and ecological integrity of the region, the government of Saskatchewan commissioned a regional environmental assessment of human-induced disturbances in the Great Sand Hills.

continued

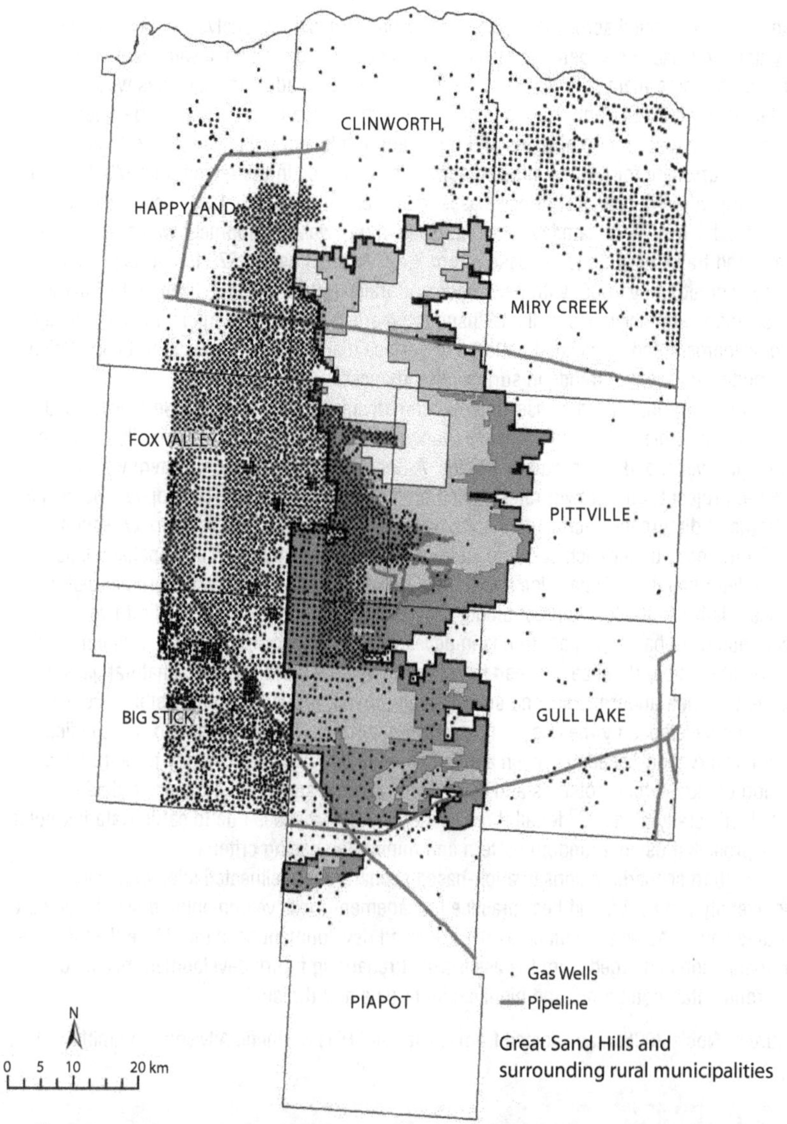

The assessment consisted of three phases: a baseline assessment, trends identification, and evaluation of alternative scenarios of human-induced surface disturbance. In the baseline assessment, concentrated biodiversity areas were delineated using a site selection algorithm, Marxan, based on collected species and habitat data. Marxan is decision-support software that uses an optimization algorithm to aid in the selection of a system of spatially cohesive sites to meet biodiversity targets. A total of 37 core biodiversity areas were identified, those with the most to lose if not managed, representing various levels of biological irreplaceability. In the second phase of the assessment, trends in surface distur-

continued

bance were identified across the landscape using retrospective analysis of aerial photography and land-use and vegetation/species databases. Rates of change were established for each of the three main sources of stress in the region—roads and trails, gas wells, and cattle watering holes. Significant statistical correlations were found to exist between gas wells and the distribution of roads and trails, and roads and trails were found to be a reasonable surrogate for human-induced surface disturbance in the region. In 1979, for example, there were 76 gas-well surface leases in the region. By 2005, 1,391 new wells had been established. Associated with the increase in gas-well development was a growth in roads and trails, which had increased from 2,497 kilometres in 1979 to approximately 3,175 kilometres by 2005, with new roads and trails 153 times more likely to be built in association with a new gas-well pad than elsewhere in the region. Annual rates of change in development and associated patterns of surface disturbance were used to build statistical models to quantify trends in surface disturbances.

In the third phase of the assessment, alternative land-use scenarios were developed based on possible futures and rates of change in human disturbance, primarily roads and trails, gas wells, and cattle watering holes. A Geographic Information System was then used to project future growth rates, based on the statistical models developed, and spatial patterns of disturbance across the landscape. Species, range, and biodiversity responses to disturbances under each scenario were modelled using statistical and spatial relationships determined in the baseline and trends analysis phases, and the patterns of disturbance relative to the locations of biodiversity hotspots were mapped. The first two scenarios were based on past trends in development and represented two variations of the future; the main difference between the scenarios was a more ambitious natural gas development agenda under the second such that all proven, probable, and possible reserves would be developed by the end of the projection period. The third scenario was predicated on a conservation-based approach and designed to conserve biodiversity through the protection of biodiversity hotspots and by further reducing surface disturbance outside the core biodiversity areas. Under all three scenarios, disturbances due to cattle watering holes were projected using a random pattern and minimum spacing criteria.

The third scenario, a conservation-based scenario that delineated sites of enhanced biodiversity protection and best-practice management for developments outside those core biodiversity areas was identified as the preferred development scenario. More than 60 recommendations emerged from the assessment regarding future development planning, environmental monitoring, and biodiversity targets and thresholds.

Sources: Noble 2008; map adapted from Great Sand Hills Scientific Advisory Committee 2007.

PREDICTING IMPACTS ON THE HUMAN ENVIRONMENT

Predicting the impacts of project development on the human environment is complex, uncertain, and rarely done well in EIA (Box 7.6). While attention is often given to impacts over which the proponent has direct control, such as employment or infrastructure development, social impacts, such as quality of life or perceptions of well-being, remain the 'orphan' of the EIA process (Burdge 2002). The role of human

Table 7.2 Sample Techniques Used for Prediction in the Biophysical Environment

Air	• dispersion modelling • box models • air quality indices • monitoring from analogues
Biological	• species population models • habitat simulation modelling • ecological risk assessment • biological assessments
Surface water	• waste load allocations • statistical models • hydrological models • water usage and allocation studies
Groundwater	• pollution source surveys • mixing models • flow and transport models • soil and groundwater vulnerability indices

Source: Based on Canter 1996.

Box 7.4 Site Water Budgets: An Example of Predicting Sub-basin Impacts on Water Quantity

Large-scale development projects, such as thermoelectric facilities and large irrigation systems, that are associated with water withdrawal or changes to the hydraulic landscape have the potential to affect surface or groundwater hydrology significantly. Thus, of central concern for such projects is the storage of water in environmental systems and the flow of water between systems. While the global circulation of water is a closed system and can be characterized by the hydrological cycle, site or regional project environments are much more open, with certain inputs and outputs that control the amount of locally stored water.

The water budget can be expressed as the change in storage DS = I – O, where I = inputs and O = outputs, and if I > O, there is an increase in water storage. Studies concerned with predicting the impacts of development projects on water storage typically focus on a particular sub-basin or geographic area from which all potential surface runoff flows through a series of streams and rivers to particular catchments or lakes. Within the water catchment, several inputs and outputs can be used to predict how changes in inputs, potentially caused by project developments and water withdrawals, might be reflected in changes in outputs. To do this, the change in storage (DS) is defined as (Pn + Rs + Rg) – (Et + Qs + Qg), where inputs are Pn = precipitation, Rs = surface run-on, Rg = groundwater seepage and outputs are Et = evapotranspiration, Qs = surface runoff, Qg = groundwater leakage. When an additional water usage is added to the equation, such as water withdrawal for thermoelectricity production, the site water budget becomes 'imbalanced'. Therefore, interruptions in the balance of the system due to project impacts can be predicted provided there is sufficient understanding of the local site hydrological regime.

Box 7.5 Dispersion Models: An Example of Predicting Plume Rise and Dispersion

Air pollution from industrial emissions can affect both human and physical environmental health. The dispersion of such pollutants is dependent on meteorological conditions (wind speed and patterns, insulation, cloud cover, precipitation patterns) and specific parameters of the pollution stack emissions (temperature, gas molecular structure, gas velocity). Pollutants exit a stack in the form of a 'plume', which resembles a moving cloud. The particular shape of the plume depends on local atmospheric conditions, which in turn determine the amount of dispersion. Predicting plume dispersion is critical to determining the ground-level concentration of the pollutants at various distances from the pollution source. Such information will allow project and environmental planners to determine appropriate stack height, severity of ground-level air pollution at various locations, the need for various mitigation measures, and compliance with emissions standards. Thus, of particular importance is predicting what is commonly referred to as the 'worst-case scenario'. Several techniques are available to predict plume dispersion and concentration; the most commonly applied is the Gaussian dispersion model. This model assumes that emissions spread outward from the source in an expanding plume according to the prevailing wind direction and that the distribution of pollution concentration decreases with increasing distance from the plume. The Gaussian model is based on a specific mathematical equation that suggests that plume dispersion, pollutant concentration, and pollutant concentrations at the surface are a function of wind speed, wind direction, and atmospheric stability. One particular example of plume analysis for predicting plume height is based on Holland's equation:

$$H = (vd) / u) (1.5 + 2.68 \times 10\text{-}3\ p\ ((Ts - Ta) / Ts)\ d)$$

where H = plume rise, v = stack gas exit velocity, d = stack diameter, u = wind speed, p = atmospheric pressure, Ts = stack gas temperature, and Ta = air temperature. Thus, plume height is determined by H = emission stack height.

Many types of Gaussian models are used to predict pollutant dispersion and ground-level concentration, including the Industrial Source Complex model, developed by the US Environmental Protection Agency, and the Atmospheric Dispersion Modelling System, developed by Cambridge Environmental Research Consultants UK.

Sources: Based on Elsom 2001; Harrop and Nixon 1999.

impact assessment in EIA is still widely debated; however, as noted at the outset of this book, 'environment' is defined here to include both the biophysical and human environments and the relations and interdependencies between them. Thus, whether human impacts are assessed separately from principal project EIAs or as part of the project EIA process, consideration of them is critical to ensuring sustainability through development.

Box 7.6 Predicting Impacts on Human Health in Canadian Northern EIA

The inclusion of health impacts in project assessment is receiving increased attention from EIA and health practitioners alike, and many health authorities, including the World Health Organization and Health Canada, have recognized the need for and benefits of addressing health in EIA. At the same time, health and other human impacts are either not considered or not given adequate treatment in project EIA. Part of the challenge is the complexity of pathways that link project development, environmental change, and human health. Recent literature on health impact assessment points to the need to focus attention not on direct cause–effect relationships for human health impact prediction but instead on the linkages between project actions and the 'determinants' of health and well-being. Determinants of health are not themselves 'health impacts'; rather, they are factors that influence or provide an indication of health and well-being—factors such as income, physical environments, health services, and social support networks. Shifting attention away from predicting impacts on human health and focusing on health determinants should simplify the complexity of pathways that connect project actions to human health impacts. Several countries, including Canada, have recently adopted health determinant frameworks for EIA to provide practitioners with advice on addressing the health implications of project development.

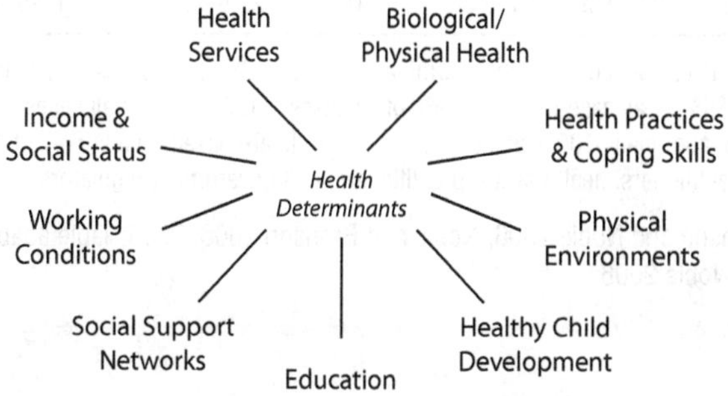

That said, recent research by Bronson and Noble (2006) and Noble and Bronson (2006; 2005) found that in practice, the consideration of health determinants in Canadian EIA, and in particular northern EIA, falls significantly short of what might be considered necessary, given the close relationship between environment, development, and the health of northern societies and cultures. While invariably considered as part of EIA, health determinants were found to be poorly integrated throughout the EIA process. When they were considered, attention focused primarily on the baseline and impact prediction phase and much less during significance determination and post-project-approval monitoring and follow-up. Further, most predicted health impacts concern physical environments, income, and education, with much less attention on issues concerning healthy child development and personal health practices and coping skills—the frequently immeasurable health determinants.

continued

Health	*Consideration in baseline and impact prediction*		*Consideration in impact significance*		*Consideration in follow-up and monitoring*	
Determinant	*Frequency*[1]	*95% CI*	*Frequency*	*95% CI*	*Frequency*	*95% CI*
Income and social status	40–59%	3.12–4.88	40–59%	2.26–3.74	1–19%	1.48–2.52
Education	40–59%	2.79–4.21	20–39%	2.51–3.49	1–19%	1.74–2.26
Physical health	20–39%	2.76–3.24	20–39%	2.26–3.74	1–19%	1.48–2.52
Health practices/ coping skills	1–19%	1.52–2.48	1–19%	1.51–2.49	1–19%	1.74–2.26
Social support networks	20–39%	1.79–3.21	20–39%	2.51–3.49	1–19%	1.48–2.52
Working conditions	20–39%	2.17–3.83	20–39%	2.26–3.74	1–19%	1.74–2.26
Physical environments	60–79%	4.05–5.95	60–79%	4.51–5.49	40–59%	2.58–3.42
Healthy child development	1–19%	1.64–2.36	1–19%	1.75–2.25	1–19%	1.74–2.26
Health services	20–39%	2.17–3.83	20–39%	2.26–3.74	1–19%	1.74–2.26

[1] The median was calculated for each determinant such that 0% of all cases = 1, 1–19% of all cases = 2, 20–39% of all cases = 3, 40–59% of all cases = 4, 60–79% of all cases = 5, 80–99% of all cases = 6, and finally 100% of all cases = 7. Results are based on a survey of 42 impact assessment practitioners, health sector practitioners, and government regulators.

Sources: Bronson and Noble 2006; Noble and Bronson 2006, 2005; Table adapted from Bronson and Noble 2006.

What to Predict

Project developments often trigger six broad areas of change that affect human systems:

- demographic change
- cultural change
- economic change
- health and social change
- biophysical change
- institutional change

Based on these changes, a number of impacts often emerge that are important to consider in project assessment (Table 7.3). That said, there is no single, comprehensive list of human impacts that should be considered in all EIAs, since the specific impacts to be predicted depend largely on the nature of the project and the local environmental setting and are defined by the scoping process.

Table 7.3 Selected Human Environmental Impacts: Examples of What to Predict

Direct economic impacts
- local and non-local employment
- labour supply and training
- wage levels
- employment demand by skill group

Indirect economic impacts
- retail expenditures
- material and service suppliers
- labour market pressures
- regional multiplier effects
- tourism

Demographic impacts
- changes in population size
- changes in population characteristics (family size, income, ethnicity)
- changes in settlement patterns

Health impacts
- quality of life (actual and perceived)
- medical standards
- worker safety (risk of death or injury)
- disease introduction and transmission
- physiological impacts (stress, worker satisfaction)
- mental and physical well-being

Housing impacts
- housing demand
- public and private housing
- house prices
- homelessness and housing problems
- density and crowding

Local service impacts
- educational services
- health services
- community services (police, fire)
- transportation services and infrastructure
- financial services

Socio-cultural impacts
- lifestyle changes (family life, seasonality of employment)
- threats to culture and belief systems
- perceived and actual risks
- social problems (crime rates, substance abuse, divorce
- community stress (conflict, integration, cohesion)
- traditional foodstuffs
- local pride and community perception
- aesthetic impacts
- gender relations

Source: Based on Glasson 2001.

Techniques

As noted above with regard to impacts on the biophysical environment, a comprehensive discussion of the full range of techniques available for predicting human environmental impacts is beyond the scope of this book. There is a long list of techniques for predicting human environmental impacts, including trends extrapolation, population multipliers, intention surveys, gravity models of spatial interaction, gaming, decision trees, traffic modelling, visibility mapping, focus group meetings, and expert judgment (Table 7.4). Boxes 7.7 and 7.8 provide examples of two different techniques for predicting human environmental impacts. Again, Morris and Therivel (2001) give the subject much more comprehensive treatment.

CHALLENGES TO IMPACT PREDICTION

In an international study of the effectiveness of environmental assessment, Sadler (1996) found that 60 per cent of the time, assessments are unsuccessful to only marginally successful in making precise, verifiable impact predictions and that 75 per

Table 7.4 Sample Techniques Used for Prediction in the Human Environment

Social	• demographic models
	• participatory mapping
	• health-based risk assessment
	• intention surveys
Economic	• economic multipliers
	• total economic productivity models
	• input-output analysis
	• Monte Carlo analysis
Cultural	• traditional knowledge
	• participatory mapping
	• community dialogues
	• analogue techniques

Source: Based on Canter 1996.

Box 7.7 Keynesian Multipliers: An Example of Predicting Economic Impacts

A commonly applied technique for predicting the broader and often indirect economic impact of project development is the **multiplier** technique. A deterministic mathematical model, multipliers represent a quantitative expression of the extent to which some initial, exogenous change is expected to generate additional effects through linkages within the economic system. The most basic of multipliers and perhaps the most commonly used for predicting local-level impacts is the **Keynesian multiplier**. The basic principle of the Keynesian multiplier is that an injection of money into a local economic system through project development will lead to an increase in the local-level monies by some multiple of the initial injection. For example, $Yr = KrJ$, where 'Yr' is the change in the level of income in the project region, 'J' is the initial income injection into the local economy as a result of project employment, and 'Kr' represents the regional income multiplier.

In principle, the commitment of a proponent to spend locally through local purchase agreements, for example, would provide an initial increase in income to the local labour force. In turn, this would generate extra income for local suppliers, who would spend that extra income in the local region. A multiplier of 1.8, for example, would mean that for each dollar injected directly into the local economy by the project, an extra 80 cents would be generated indirectly (Morris and Therivel 2001).

However, the local project environment is not isolated, and thus economic leakages reduce the multiplier effect. For example, some of the additional income is lost to taxation, individuals may choose to save their extra income rather than spend it, or individuals may choose to spend that extra income in other communities or regions outside the local environment. Thus, a multiplier for predicting local economic impacts of an initial project economic investment might more closely resemble: $Yr = [1 / 1 - (1\text{-}s)(1\text{-}tx\text{-}u)(1\text{-}m)(1\text{-}tl)] \times J$, where 's' is the proportion of additional income generated that is lost to the economy through personal savings, 'tx' is the proportion of additional income that is lost to taxes on income, 'u' is attributed to the reduction in government support and transfers as local employment increases, 'm' is the proportion of additional income spent on goods and serv-

continued

ices outside the region, and 'tl' is the proportion of additional income that is lost to taxes on goods and services purchased.

Typically, Keynesian multipliers are limited in EIA to the prediction of indirect economic impacts during peak and trough periods of project construction or operation—that is, best-case and worst-case scenarios.

Box 7.8 Decibel Ratings: An Example of Predicting Cumulative Noise Impacts

Noise is often an impact generated by large-scale projects such as airports, highways, railways, or heavy industries. The intensity of sound, while often expressed as 'loudness', is actually measured in decibels (dB). The range of audible sounds is typically from 0 dB to 140 dB. However, because sound level is logarithmic in nature, adding two sounds or doubling the intensity of a sound does not result in a doubling of the decibel rating. For example, adding two identical sounds (e.g., two low-level aircraft passing overhead at once) does not double the decibel rating but increases it by only 3 dB. When two sources of sound that are at different decibel ratings are added together (e.g., sound source 1 = 59 dB; sound source 2 = 53 dB), the graph below can be used to predict the cumulative noise rating (e.g., 59 dB – 53 dB = 6 dB difference; therefore, according to the graph, add 1 dB to the higher level sound = total 60 dB). To add three sounds each of 50 dB, we add 50 dB + 50 dB = 53 dB, then 50 dB + 53 dB = 54.75; the difference between 50 dB and 53 dB is 3 dB, so according to the graph we add 1.75 dB to the higher rating.

Sound level (dB)	*Example*	*Sound level (dB)*	*Example*
140	pain threshold for human hearing	70	busy street corner
120	pneumatic drill	60	busy office place
110	loud car horn at one metre	30	bedroom at night
105	jet flying overhead at 250 metres	10	normal breathing
90	inside a subway car	0	threshold of hearing

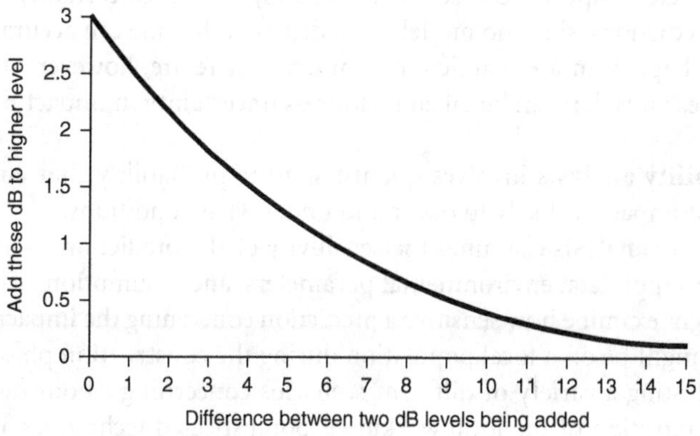

continued

This is a rather simplistic approach to predicting noise impacts. In practice, several additional elements have to be factored into the equation, such as the level of other sounds in the environment, meteorological effects such as average wind direction and speed, local topography, other physical obstructions between the noise source and the noise receptor, and the sensitivity of the noise receptor. In addition, the impacts of noise are not limited to the human environment or to human assessments of loudness; rather, as illustrated by low-level military flight training in Labrador (Box 3.3), impacts on wildlife and the startle effect of noise are also important considerations.

Sources: Based on Harrop and Nixon 1999; Morris and Therivel 2001.

cent of the time, assessments fail to indicate confidence levels of the data used for impact predictions.

In international EIA experience, little has changed since Bisset and Tomlinson (1988) noted that 697 of a total 791 predicted impacts across a survey of four UK impact assessments could not be confirmed, partly because of the vagueness of impact predictions. Similarly, Bailey, Hobbs, and Saunders (1992), in an environmental audit of waterway developments in Western Australia, found that 90 per cent of impact predictions did not indicate a specific time scale in which the predicted impact was expected to occur. This is confirmed by Morrison-Saunders and Bailey (1999), who more recently analyzed six Australian cases and found little evidence of impact quantification or precision in prediction, with most predictions being vague and qualitative in nature. In the Canadian experience, Beanlands and Duinker (1983) examined the scientific quality of EIA in Canada from an ecological perspective. Their analysis suggested the limited value of impact predictions for decision-making and subsequent testing and replication. To address this issue, EIA practitioners need to recognize that uncertainty is inevitable and that impact predictions must be stated in such a way that they can be verified.

Addressing Uncertainty in Impact Prediction

The question that must be asked for any impact prediction is: 'How certain are we that the predicted impact is correct or will actually occur?' Uncertainty is inevitable in impact predictions, since no model or predictive technique can accurately predict what might happen in a dynamic environment. There are, however, at least three principal measures that can be taken to address uncertainty in impact prediction:

1. **Probability analysis** involves quantifying the probability that an impact or range of impacts is likely to occur and under what conditions.
2. **Sensitivity analysis** examines the sensitivity of the prediction to minor variations in input data, environmental parameters, and assumptions. For example, one might examine how sensitive a prediction concerning the impacts of worker influx might be on a local population during the construction phase of a project by testing a variety of different scenarios concerning labour demands and the composition of the local workforce. Sophisticated techniques are available

for evaluating the sensitivity of impact predictions, such as **Monte Carlo analysis**, a mathematical technique using random samples, while others simply involve providing a range of impact predictions for a particular impact indicator based on minor variations in the data and parameters on which those predictions are based.

3. **Confirmatory analysis** is used to test for uncertainty regarding the predictive technique itself and to ensure that the predicted impact is not solely a product of the particular technique used. In other words, the objective is to determine the extent to which different techniques provide similar predicted outcomes.

Verifiable Impact Predictions

A characteristic critical to the value of impact predictions is that they be stated in such a way that they can actually be followed up and verified. This can be accomplished by one of two methods depending on the nature and availability of baseline data.

Hypotheses-based approaches. When impacts are predicted in environmental assessment, they should, where at all possible, be based on rigorous and falsifiable null-impact hypotheses stating the relevant affected variables, impact magnitude, spatial and temporal extent, probability of occurrence, significance, and associated confidence intervals. Ringold et al. (1996) suggest that impact predictions should be couched in quantitative terms, should identify the spatial and temporal characteristics of interest, and should state the Type I and Type II probabilities. This is consistent with recommendations of the panel reviewing the recent Voisey's Bay nickel mine project proposal for northern Labrador, which suggested that a hypothesis-driven approach is the preferred formula for impact prediction (Voisey's Bay Panel 1999). Such an approach to impact prediction allows practitioners to measure variables and either accept or reject impact predictions and environmental change within specified confidence limits. However, it is important to note that a hypothesis-driven approach to impact prediction does not supplant the role of professional judgment when considering what constitutes a significant or acceptable deviation from the null and that it inevitably relies on the availability and quality of data.

A hypothesis-driven approach requires that assumptions underpinning impact predictions be clearly stated, including any exogenous factors associated with the impact prediction. Munro, Bryant, and Matte-Baker (1986), for example, identify a range of probability and confidence intervals associated with impact predictions (Table 7.5). The lower the associated confidence level, the greater the likelihood that effects monitoring will become important as the project proceeds.

Threshold-based predictions. Impact prediction typically involves the use of some model either mathematically or conceptually representing the biophysical or socio-economic environment. Uncertainties emerge, contributing to inaccurate impact predictions, because the ultimate purpose of such models is to simplify the actual system and inherently, they cannot capture the multiplicity of factors interacting and affecting the natural system (De Jongh 1988). In any environmental or socio-economic system, several different processes are often involved that may affect the variable or environmental component of concern. Thus, impact predictions often turn out to be inaccurate because of the mix of assumptions that normally have to be made and the multiplicity of exogenous factors involved (Mitchell 2002, 56). First, projects are almost always modified during the development phase, thereby making initial impact predictions less valuable from a follow-up and learning perspective. Second, given the nature of constantly changing and often unpredictable human environmental systems, the environmental impacts of development can rarely be predicted with any degree of certainty. This is particularly true in the social environment, where humans often react and adapt in anticipation of project-induced change.

However, not all environmental effects need to be predicted precisely; in many cases, particularly when baseline data are inadequate, it is recognized in the assessment that outcomes should not exceed specified threshold levels, in which case management practices, the setting of targets, and the determination of threshold levels become the focus of attention (Storey and Noble 2004). It is often the case, for example, that a threshold or particular target is set and a monitoring framework is established to ensure that negative impacts do not exceed certain thresholds and that positive impacts meet specified expectations. Where hypothesis-based approaches are not suitable, **threshold-based predictions** may rely on previous experience with similar projects, similar types of impacts in different environments, or regulatory standards. For example, it is known that certain levels of noise can have a negative effect on human hearing and that certain project construction activities (such as heavy machinery or blasting) generate particular levels of noise. One approach would be to compare the level of cumulative noise to specified thresholds for human hearing in order to predict the impact of project construction activities.

When comparative examples or regulatory standards are not readily available, an alternative approach is to base impact predictions on a desirable level or maximum level of change. **Maximum allowable effects levels**, for example, reflect an approach to impact prediction in which the impact is stated in the form of a hypothesis such as 'project impact *i* will not exceed a particular threshold or desired effects level for a particular impact indicator *j*.' This approach was used in the Hibernia offshore oil project for predicting potential project impacts on the biophysical environment, and it is particularly useful for managing project outcomes and meeting desired sustainability objectives.

Table 7.5 Impact Predictions and Confidence Limits

Confidence levels	*Data characteristics*	*Knowledge*	*Permitted approach*
high/factual (95%)	reliable	proven cause–effect relationship	statistical prediction
fairly high	sufficient	evidence for hypotheses	quantitative simulation
fairly low	insufficient	postulated linkages	conceptual modelling
low/intuitive (50%)	absent or unreliable	speculation	professional opinion

Source: Munro, Bryant, and Matte-Baker 1986.

KEY TERMS

accuracy
backcasting
balance model
confirmatory analysis
Delphi technique
deterministic model
gravity model
intention survey
Keynesian multiplier
maximum allowable effects level
mechanistic model
Monte Carlo analysis
multiplier
precision
probability analysis
sensitivity analysis
statistical model
stochastic model
threshold-based prediction

STUDY QUESTIONS AND EXERCISES

1. Discuss the relationship between accuracy and precision in impact prediction.
2. What are some of the challenges to making impact predictions concerning biophysical and social change?
3. What are the advantages and disadvantages of analogue techniques for impact prediction?
4. Expert judgment is perhaps the most common predictive element in most EIAs. Discuss the advantages and limitations of relying on expert judgment for impact prediction.
5. Consider a condition with the following baseline noise environment:
   - average daily road traffic 55 dB
   - average daily rail traffic 48 dB
   - average daily air traffic 58 dB
   - average noise from factories 62 dB
6. What would be the total noise impact of a proposal to develop a heavy manufacturing plant in the region with equipment operating at average daily noise levels of 68 dB? Would you consider this a significant impact? Explain. How might you manage such an impact?
7. Obtain a completed project EIS from your local library or government registry, or access one on-line. Scan the document for stated impact predictions and examine how these predictions are stated. Are the statements based on verifiable hypotheses? Are thresholds or maximum allowable effects levels stated? Given the nature of the predictions and the way in which they are stated, do you think that they can be followed up and verified? Compare your findings with those of others.
8. Obtain a completed project EIS from your local library or government registry, or access one on-line. Scan the document and generate a list of the techniques used to predict, describe, or assess impacts on the biophysical and human environments. Compare your results with those of others. Is there a common set of techniques that emerge for various environmental components, such as water quality, air quality, or employment?

REFERENCES

Bailey, J., V. Hobbs, and A. Saunders. 1992. 'Environmental auditing: Artificial waterway developments in Western Australia'. *Journal of Environmental Management* 34: 1–13.

Beanlands, G.E., and P.N. Duinker. 1983. 'Lessons from a decade of offshore environmental impact assessment'. *Ocean Management* 9 (3/4): 157–75.

Bisset, R., and P. Tomlinson. 1988. 'Monitoring and auditing of impacts'. In P. Wathern, Ed., *Environmental Impact Assessment: Theory and Practice*, 117–26. London: Unwin Hyman.

Bronson, J., and B.F. Noble. 2006. 'Health determinants in Canadian northern environmental impact assessment'. *Polar Record* 42 (223): 315–24.

Buckley, R. 1991. 'Auditing the precision and accuracy of environmental impact predictions in Australia'. *Environmental Monitoring and Assessment* 18: 1–23.

Burdge, R. 2002. 'Why is social impact assessment the orphan of the assessment process?' *Impact Assessment and Project Appraisal* 20 (1): 3–11.

Canter, L.W. 1996. *Environmental Impact Assessment*. 2nd edn. New York: McGraw Hill.

Dale, V., and M.R. English, Eds. 1999. *Tools to Aid Environmental Decision-Making*. New York: Springer.

Dalkey, N. 1969. 'An experimental study of group opinion: The Delphi method'. *Futures* 1: 408–20.

De Jongh, P. 1988. 'Uncertainty in EIA'. In P. Wathern, Ed., *Environmental Impact Assessment: Theory and Practice*. London: Unwin Hyman.

Duinker, P.N., and G.E. Baskerville. 1986. 'The significance of environmental impacts: An exploration of the concept'. *Environmental Management* 10 (1): 1–10.

Duinker, P.N., and L. Greig. 2007. 'Scenario analysis in environmental impact assessment: Improving explorations of the future'. *Environmental Impact Assessment Review* 27 (3): 206–19

Elsom, D.E. 2001. 'Air quality and climate'. In P. Morris and R. Therivel, Eds, *Methods of Environmental Impact Assessment*, 2nd edn. London: Taylor and Francis Group.

Glasson, J., R. Therivel, and A. Chadwick. 1999. *Introduction to Environmental Impact Assessment: Principles and Procedures, Process, Practice and Prospects*. 2nd edn. London: University College London Press.

Great Sand Hills Scientific Advisory Committee. 2007. *Great Sand Hills Regional Environmental Study*. Regina: Canada Plains Research Centre.

Greig, L., K. Pawley, and P. Duinker. 2002. *Alternative Scenarios for Future Development: An Aid to Cumulative Effects Assessment*. Gatineau, QC: Canadian Environmental Assessment Agency.

Harrop, D.O., and A.J. Nixon. 1999. *Environmental Assessment in Practice*. Routledge Environmental Management Series. London: Routledge.

Kahn, H., and A. Wiener. 1967. *The Year 2000*. New York: MacMillan.

Mitchell, B. 2002. *Resource and Environmental Management*. 2nd edn. New York: Prentice-Hall.

Morris, P., and R. Therivel, Eds. 2001. *Methods of Environmental Impact Assessment*. 2nd edn. London: Taylor and Francis Group.

Morrison-Saunders, A., and J. Bailey. 1999. 'Exploring the EIA/environmental management relationship'. *Environmental Management* 24 (3): 281–95.

Munro, D., T. Bryant, and A. Matte-Baker. 1986. *Learning from Experience: A State of the Art Review of Environmental Impact Assessment Audits*. Ottawa: Canadian Environmental Assessment Research Council.

Noble, B.F. 2008. 'Strategic approaches to regional cumulative effects assessment: A case study of the Great Sand Hills, Canada'. *Impact Assessment and Project Appraisal* 26 (2): 78–90.

Noble, B.F., and J. Bronson. 2005. 'Integrating human health in environmental impact assessment: Case studies of Canada's northern mining resource sector'. *Arctic* 58 (4): 395–405.

———. 2006. 'Practitioner survey of the state of health integration in environmental assessment: The case of northern Canada'. *Environmental Impact Assessment Review* 26 (4): 410–24.

Ringold, P.R., et al. 1996. 'Adaptive monitoring design for ecosystem management'. *Ecological Applications* 6 (3): 745–7.

Sadar, H. 1996. *Environmental Impact Assessment.* 2nd edn. Ottawa: Carleton University Press.

Sadler, B. 1996. *Environmental Assessment in a Changing World: Evaluating Practice to Improve Performance.* Final report of the International Study of the Effectiveness of Environmental Assessment. Fargo, ND: IAIA.

Storey, K. 1995. 'Managing the impacts of Hibernia: A mid-term report'. In B. Mitchell, Ed., *Resource and Environmental Management in Canada*, 2nd edn., 310–34. Toronto: Oxford University Press.

Storey, K., and P. Jones. 2003. 'Social impact assessment, impact management and follow-up: A case study of the construction of the Hibernia offshore platform'. *Impact Assessment and Project Appraisal* 21 (2): 99–107.

Storey, K., and B. Noble. 2004. *Toward Increasing the Utility of Follow-up in Canadian EA: A Review of Concepts, Requirements and Experience.* Report prepared for the Canadian Environmental Assessment Agency. Gatineau, QC: CEAA.

Voisey's Bay Mine and Mill Environmental Assessment Panel. 1999. *Report on the Proposed Voisey's Bay Mine and Mill Project / Environmental Assessment Panel.* Hull, QC: Canadian Environmental Assessment Agency.

Woudenberg, F. 1991. 'An evaluation of Delphi'. *Technological Forecasting and Social Change* 40: 131–50.

CHAPTER 8

Determining Impact Significance

IMPACT SIGNIFICANCE

Determining the significance of environmental effects is one of the most critical components of the EIA process but among the most complex of all EIA activities. Several attempts have been made to define significance in relation to environmental effects; however, there is still no accepted interpretation of significance or standard methods for significance determination (Table 8.1). Part of the complexity surrounding significance is that there is no generic standard that defines significance, there will always be site-specific issues to take into consideration, and there is also considerable variation in the types of activities and impacts being assessed as significant in any given EIA.

Table 8.1 Interpretations of 'Significance'

Canter and Canty (1993)	Significance can be considered on three levels: (1) significant and not mitigatible; (2) significant but mitigatible; and (3) insignificant. Significance is sometimes based on professional judgment, executive authority, the importance of the project/issue, sensitivity of the project/issue, and context, or by the controversy raised.
US Council on Environmental Quality (1987)	The United States NEPA requires significance to be determined within the framework of context and intensity. Context—the significance of an action must be analyzed in several contexts such as society as a whole, the affected region, the affected interests, and the locality. Intensity—refers to the severity of impact.
Duinker and Beanlands (1986)	Significance of environmental impacts is centred on the effects of human activities and involves a value judgment by society of the significance or importance of these effects. Such judgments, often based on social and economic criteria, reflect the political reality of impact assessment in which significance is translated into public acceptability and desirability.
Haug et al. (1984)	Determining significance is ultimately a judgment call. The significance of a particular issue is determined by a threshold of concern, a priority of that concern, and a probability that a potential environmental impact may cross the threshold of concern.
Sadler (1996)	The evaluation of significance is subjective, contingent upon values, and dependent upon the environmental and community context. Scientific disciplinary and professional perspectives frame evaluations of significance. Scientists therefore evaluate significance differently from one another and from local communities.

Despite the lack of a single definition of significance, Sippe (1999) recognizes that some degree of commonality exists across the various interpretations of the concept—namely, significance:

- is an informed judgment;
- depends, in part, on the characteristics of anticipated environmental impacts and the nature of the receiving environmental components;
- is both biophysical and socio-economic in context;
- involves some level of change, including cumulative change, which is perceived to be acceptable to the interested parties.

Determination of impact significance then essentially involves making judgments about the importance of environmental effects. Significance reflects the degree of importance placed on the effects in question and consists of two components:

- the significance of predicted project effects;
- the significance of predicted effects following impact management or mitigative measures.

Effects associated with the latter are typically referred to as **residual effects**, or impacts that remain after all management and mitigation measures have been implemented.

The determination of significance begins at the outset of the EIA process when a decision is being made as to whether the proposal requires a formal assessment and extends throughout the scoping, prediction, mitigation, and follow-up stages (Table 8.2). The information and characteristics used to define and support significance determination during the early stages of screening and scoping differ from the conceptualization and determination of significance during impact prediction, evaluation, and verification of impact mitigation effectiveness (e.g., see Wood and Becker 2004; Sadler 1996; Westman 1985). It is possible, explains Lawrence (2005), to apply significance thresholds and criteria to determine whether an EIA is required and to focus or scope the EIA process. Following screening and scoping, significance determination can involve applying different types of quantitative thresholds and criteria to individual and cumulative effects, taking into account context, stressor characteristics, public interest, uncertainty, and the characteristics of the receiving environment.

DETERMINING SIGNIFICANCE

Notwithstanding the importance of determining significance, there is little consistent guidance for practitioners. Generally speaking, effects are likely to be considered significant if they are:

- adverse;
- intensive in concentration or associated with significant levels of change;
- associated with a high degree of probability;
- frequent and long-lasting;
- likely to occur at a broad spatial scale;

Table 8.2 Interpretations of Significance in the EIA Process

EA Activity	*Significance Interpretation*
Screening	If (and what) EIA requirements are to be applied Criteria and procedures for making screening decisions
Scoping	Alternatives that are reasonable and criteria for comparing them Analysis of boundaries and components to focus on Public and agency issues Proposal characteristics most likely to induce significant effects Proposal characteristics that warrant mitigation and monitoring
Baseline analysis	Criteria for determining environmental significance and sensitivity Choice of valued ecosystem components
Impact analysis and interpretation	Potential impacts to analyze and at what level of detail Impact magnitude and impact significance criteria and thresholds Impact acceptability and significance interpretations
Cumulative effects analysis	Cumulative effects criteria and thresholds Acceptability of cumulative effects Cumulative effects significance interpretations
Decision-making	Proposal acceptability and compliance with standards, policies, etc. Basis for decision-making
Mitigation	Impacts that warrant mitigation, compensation, or benefits Choice of measures and significance of residual impacts
Monitoring and auditing	Impacts that should be monitored and choice of monitoring methods Effectiveness of mitigation and monitoring

Source: Based on Lawrence 2000.

- irreversible;
- associated with cumulative change;
- going to detract from the sustainability of environmental and socio-economic systems;
- likely to affect ecological functions or exceed **assimilative capacity** of the environment;
- associated with variables of societal importance and public concern;
- not in compliance with existing standards or regulations;
- likely to exceed desired levels of change.

Impact significance, then, is a function of the characteristics of the environmental effect or impact and the importance or value attached to the affected component (Figure 8.1).

Effect and Impact Characteristics

For each prediction, several questions concerning the characteristics of environmental effects and impacts are typically asked (see Lawrence 2004; Canter 1996;

Impact significance	=	impact characteristic (e.g., duration, extent)	×	impact importance (e.g., value of receptor)

Figure 8.1 Definition of impact significance.

Wood 1995). The specific nature of these questions varies from project to project but usually involves the identification of 'major', 'moderate', 'minor', and 'negligible' effects based on several characteristics (Table 8.3). Many of these characteristics were discussed in Chapter 3 and are reviewed only briefly here, with particular attention to other characteristics not previously introduced.

Adverse effect. A first step in determining significance is to determine whether effects are likely to be adverse following effects mitigation. Some major factors that should be used to determine whether environmental effects are adverse are described by FEARO (1994) and summarized in Table 8.4. The most common way of determining whether effects are adverse is to compare the quality of the environment with and without the project, using as variables the relevant environmental components identified in Table 8.4. This does, however, imply that baseline data are available or can be established for each of the variables concerned. The concept of

Table 8.3 Selected Impact Significance Criteria for the Ekati Diamond Mine EIA

	Type of environmental component		
Impact significance	*Physical*	*Biological*	*Socio-economic*
Major	Parameter affected within most of ecozone for several decades	Whole stock or population of ecozone affected over several generations	Whole population of region affected over several generations
Moderate	Parameter affected within most of ecoregion for one or more decades	Portion of population of ecoregion affected over one or more generations	Community affected over one or more generations
Minor	Parameter affected within most of ecoregion during less than one decade	A specific group of individuals within an ecosystem affected during less than one generation	A specific group of individuals within a community affected during less than one generation
Negligible	Parameter affected within some part of ecoregion for a short time period	A specific group of individuals within an ecosregion affected for a short time period	A specific group of individuals within a community affected for a short time period

Source: Based on BHPB 1998.

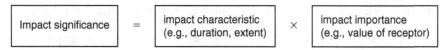

Table 8.4 Selected Factors in Determining Adverse Environmental Effects

Changes in the environment	*Human impacts resulting from changes*
Negative effects on the health of biota, including plants, animals, and fish	Negative effects on human health, well-being, or quality of life
Threat to rare or endangered species	Increase in unemployment or shrinkage in the economy
Reduction in species diversity or disruption of food webs	Reduction of the quality or quantity of recreational opportunities or amenities
Loss or damage to habitats, including habitat fragmentation	Detrimental change in the current use of lands and resources for traditional purposes by Aboriginal persons
Discharges or release of persistent and/or toxic chemicals, microbiological agents, nutrients (e.g., nitrogen, phosphorus), radiation, or thermal energy (e.g., cooling waste water)	Negative effects on historical, archaeological, paleontological, or architectural resources
Population declines, particularly in top predator, large, or long-lived species	Decreased aesthetic appeal or changes in visual amenities (e.g., views)
The removal of resource materials (e.g., peat coal) from the environment	Loss or damage to commercial species or resources
Transformation of natural landscapes	Foreclosure of future resource use or production
Obstruction of migration or passage of wildlife	
Negative effects on the quality and/or quantity of the biophysical environment (e.g., surface water, groundwater, soil, land, and air)	

Source: FEARO 1994.

significance illustrated by the FEARO guidelines is closely related to the concepts of 'adverse' and 'likely', and a three-step framework sets this relationship (Baker and Rapaport 2005):

- deciding whether the environmental effects are adverse;
- deciding whether the adverse effects are significant (based on impact characteristics and the characteristics of the receiving environment);
- deciding whether the significant adverse effects are likely (based on probability and uncertainty).

Magnitude. The degree or amount of change associated with an adverse environmental effect, magnitude is usually not considered high if a major adverse effect can be mitigated. It is important to emphasize that magnitude itself does not always equate directly with significance. For example, a large increase in mercury levels in a local water supply reservoir may not be considered significant if the mercury levels are still within the quality standards for drinking water. The significance of an effect, in terms of its magnitude, is typically assessed in comparison to deviation from a pre-project baseline condition or from an alternative predetermined measurement point. For example, in the Jack Pine mine EIS, impact magnitude measures for noise, groundwater hydrology, surface water hydrology, and wildlife health were determined on the basis of specified measured baseline conditions (Table 8.5).

Probability. If an environmental effect or impact is not likely to happen, then it may not be significant. The determination of likelihood is based on two criteria: (1) the probability of occurrence and (2) scientific certainty. In practice, the likelihood of an attribute of significance is often rated on a scale of 'none' (0 per cent chance), 'low' (< 25 per cent chance), 'moderate' (25–75 per cent chance), and 'high' (> 75 per cent chance). While sometimes based on **statistical significance**, this is only one means of determining probability (Table 8.6).

Duration and frequency. Significance may be based on the duration of the environmental effect or impact—short-term (one to five years after project completion), medium-term (six to 15 years after completion), or long-term (more than 15 years). This is, however, a highly subjective process and should be based on similar activities and impacts from comparable projects when possible. It is also important to consider whether the impact is continuous, delayed, or immediate. An impact likely to be delayed for 10 years after project completion may or may not be considered as significant as an immediate and continuous impact.

Spatial extent. As depicted in Table 8.3, localized or contained effects may not be considered as significant as effects or impacts observed or experienced at locations far

Table 8.5 Selected Magnitude Criteria for the Jack Pine Mine EIA

Factor	*Impact magnitude criteria and indices*
Noise	Negligible (0): no projected increase in ambient noise levels Low (+5): increased noise levels do exist but do not exceed the nighttime noise criteria Moderate (+10): increased noise levels exceed to the nighttime criteria High (+15): increased noise levels exceed the daytime noise criteria
Groundwater hydrology	Negligible (0): no change from the baseline case Low (+5): near (slightly above) the baseline case Moderate (+10): above the baseline case High (+15): substantially above the baseline case
Surface water hydrology	Negligible (0): < 5 per cent change Low (+5): 5–10 per cent change Moderate (+10): 10–30 per cent change High (+15): > 30 per cent change
Wildlife health	Negligible (0): no appreciable increase in hazard compared to baseline Low (+5): possible small increase in hazard compared to baseline Moderate (+10): possible moderate increase in hazard compared to baseline High (+15): substantial increase in hazard compared to baseline

Source: Shell Canada Limited 2002.

Table 8.6 Impact Probability Classifications

High	Previous research, knowledge, or experience indicates that the environmental component *has experienced* the same impact from activities of similar types of projects.
Moderate	Previous research, knowledge, or experience indicates that the environmental component *may have experienced* the same impact from activities of similar types of projects.
Low	Previous research, knowledge, or experience indicates there is a *small likelihood* that the environmental component has experienced the same impact from activities of similar types of projects.
Unknown	There is *insufficient research, knowledge, or experience* to indicate whether the environmental component has experienced the same impact from activities of similar types of projects.

Source: Based on BHPB 1998.

removed from the project, such as **acid mine drainage** or atmospheric emissions. Spatial scale provides a frame of reference for EIA and is an important contextual factor in interpreting significance (see Irwin and Rodes 1992). The challenge, however, is that if large boundaries are defined in determining the significance of a proposed activity, then only a superficial assessment of significance may be possible, and uncertainty will increase. If the boundaries are small, then a more detailed examination may be feasible, but an understanding of the broader context and significance of the proposal may be sacrificed. The spatial scale of the proposed project assessment and of the effects should be considered in significance determination, alongside the nature of the proposal, the characteristics of anticipated effects, and whether cumulative or incremental effects are deemed important to consider within the spatial boundary.

Mitigable effects. Environmental effects are typically considered less significant if they can be mitigated. One source of particular controversy between proponents and regulators, however, is the notion and role of impact mitigation and mitigated FONSIs—Findings of Non-Significant Impacts. A mitigated FONSI refers to a proposed action that has incorporated mitigation measures to reduce any significant negative effects to insignificant ones (Canter 1996). Impact mitigation is inherent in all aspects of EIA systems, and an important issue in significance determination is whether a significant impact can, and will, be sufficiently mitigated. It is common practice under the US NEPA system, for example, for proponents to prepare and submit project proposals and environmental management plans identifying impact mitigation measures that result in mitigated FONSIs. Tinker et al. (2005) explain that a fundamental problem with this approach, unless the environmental management plan is very precise about specific mitigation measures, is that it is not possible to create a valid condition requiring the development to comply with the proposed mitigation. It should not be assumed that conformity with proposed mitigation rules out the need for EIA. In most cases, mitigation is a series of non-binding proposals in a project proposal or preliminary project or impact management plan (Morrison-Saunders et al. 2001). Byron (2000) reports that mitigation measures proposed give no indication as to whether they will actually be implemented or their effectiveness in mitigating adverse effects. Mitigation measures should not be ignored when making decisions about the likely significant effects of proposed development, but they should not form the lead criterion in the decision as to whether an EIA is required. In cases where mitigation measures are considered in the significance determination process, a decision-maker should consider the likelihood that such mitigation will occur, factors to ensure implementation, and proven effectiveness of the mitigation measure (Ross, Morrison-Saunders, and Marshall 2006).

Standards and regulations. Predicted effects or impacts following mitigation are often compared against environmental standards or regulations—in essence, specified thresholds. A threshold involves a clearly defined performance level, usually applied to distinguish between significant and insignificant effects. Standards and regulations are perhaps the most common approach to thresholds for assessing impact significance. Environmental impacts within specified standards or regulations are often deemed to be insignificant in comparison to impacts that exceed regulations or, for example, those that attain maximum allowable concentrations. Many jurisdictions have standards for drinking water quality or levels of industry emissions. Lawrence (2004) identifies at least three major types of standards or regulatory thresholds:

- *exclusionary:* leads to automatic rejection of a proposal;
- *mandatory:* leads to a mandatory finding of significance;
- *probable:* normally significant but subject to confirmation.

However, with the exception of California and Australia, significance thresholds are not normally included in EIA requirements.

Cumulative effects. Environmental effects that are cumulative in nature—that is, they add to or interact with an already existing adverse impact to detract additively or synergistically from environmental quality—are more likely to be deemed significant compared to impacts that do not affect already affected environmental components. Significance cannot be avoided by terming an action 'temporary' or by breaking it down into small component parts (Canter 1996). Moreover, significance cannot be defined taking into account a proposed development in isolation from other existing and proposed activities in the region. In practice, significance determination in EIA typically focuses on evaluating the effects of a proposed action and the linear, causal effects, direct and indirect, on particular environmental components at the site of the proposed action. Although often individually insignificant, a combined number of small-scale alterations (e.g., a program of development) can lead to significant environmental change. Accordingly, a comprehensive approach to determining the significance of individual proposals that considers cumulative and induced effects is required.

The term 'cumulative environmental effects' is generally used to describe the phenomenon of environmental change resulting from numerous, small-scale alterations, whether they are due to the activities of a single development or the combined effects of multiple developments over space and time. Significance determination should involve considering whether the proposed action is related to other actions that may result in individually insignificant but cumulatively significant effects or impacts and whether the proposed action may establish a precedent for future actions or represent a decision, in principle, about future actions. In other words, a piecemeal approach to assessing proposed developments might circumvent the identification and management of potentially significant cumulative effects. Questions concerning the significance of project impacts should be addressed in light of past, existing, and future actions.

Sustainability criteria. If the underlying objective of EIA is to contribute to the sustainable development of the environment, then an important question to ask in determining significance is whether the proposal will make an overall positive contribution to sustainability. While the definitions and interpretations of sustainability vary considerably, it is generally acknowledged that sustainability includes the desire to maintain, over an indefinite future, necessary and desired attributes of ecological and socio-economic environments that are necessary for system functioning and integrity (Deakin, Curwell, and Lombari 2002). Thus, an alternative approach to determining impact significance is to ask whether the project's impacts make a positive contribution to or detract from sustainability. Gibson (2001), for example, identifies several generic sustainability-based questions for evaluating the significance of environmental impacts, including:

- Could the effect add to stress that might undermine ecological integrity or damage life-support functions?
- Could the effect contribute to ecological rehabilitation or otherwise reduce stress on environmental components?
- Could the effect increase equity in the provision of material security, including that of present and future generations?
- Could the effect provide more opportunities for economic well-being while reducing material and energy demands?

In the case of the Voisey's Bay mine project, introduced in Chapter 1, the 'sustainability of resources', defined as the capacity of an affected resource to meet present and future needs (Voisey's Bay Nickel Company 1997), was used as a factor to assist in determining the significance of residual impacts based on such concepts as ecosystem integrity, carrying capacity, and assimilative capacity (Table 8.7).

Impact Importance

An important aspect of interpreting significance is the consideration of context—that significance should be assessed relative to society, the affected region, the affected interests, and the locality (Canter 1996). The concept of significance in EIA was defined initially by lists of actions regarded as significantly affecting the environment and the identification of suitable assessment thresholds (US Council on Environmental Quality 1973). While these processes considered the nature and scale of effects in significance determination, the context was disregarded (Benson 2004). Significance determination is highly context-sensitive (Kjellerup 1999). What is considered significant is thus highly dependent on a number of contextual factors in addition to the characteristics of the effect itself. In other words, the importance of a particular environmental effect is related to ecological and societal values about the

Table 8.7 Sustainability Criteria Used in the Voisey's Bay EIS for Significance Determination

Sustainability rating	*Sustainable use of renewable resources criteria*
High	Previous research/experience indicates that the environmental effect on the VEC would not reduce biodiversity or the capacity of resources to meet present and future needs.
Moderate	Previous research/experience indicates that the environmental effect on the VEC may, to a certain extent, reduce biodiversity or the capacity of resources to meet present and future needs.
Low	Previous research/experience indicates that the environmental effect on the VEC would reduce biodiversity or the capacity of resources to meet present and future needs.
Nil	Previous research/experience indicates that the environmental effect on the VEC would eliminate biodiversity or the capacity of resources to meet present and future needs.
Unknown	There is insufficient research/experience to indicate whether the environmental effect on the VEC would reduce biodiversity or the capacity of resources to meet present and future needs.

Source: Voisey's Bay Nickel Company 1997.

affected components. The importance or value of the affected components is usually based on societal, ecological, economic, political, public, or regulatory concern.

Ecological value. From an ecological perspective, potential effects on rare or threatened plants, animals, and their habitat, or on particularly vulnerable species or components of the natural environment that are essential to ecological functioning, are often deemed to be significant impacts. The effects of development in highly sensitive areas or on rare or endangered species or features of the environment, for example, are more likely to be considered significant than the same effects on a resilient or 'common' landscape (Table 8.8). In the absence of legal or policy instruments, however, or existing land-use designations that clearly identify such features a priori, interpretations of significance based on the sensitivity or uniqueness of the receiving environment should be supported by baseline information, trends analysis, and widespread public consultation or with reference to specified thresholds, criteria, or desired outcomes and objectives for the receiving environment or landscape.

Societal value. Project effects on the biophysical environment often affect factors of importance to society and human life, as well as biophysical components that are valued for aesthetic or sentimental reasons. Societal values that can be affected include:

- human health and safety;
- potential loss of green space;
- recreational value;
- demands on public resources;
- demands for infrastructure and services;
- demographic effects.

Table 8.8 Example Significance Determination for Sensitivity of Landscape Receptors

High sensitivity	Key characteristics and features that contribute to the distinctiveness and character of the landscape. Designated landscapes (e.g., national parks) and landscapes identified as having low capacity to accommodate the proposed form of change.
Medium sensitivity	Other characteristics or features of the landscape that contribute to the character of the landscape locally. Locally valued landscapes that are not designated. Landscapes identified as having some tolerance of the proposed change subject to design and mitigation, etc.
Low sensitivity	Landscape characteristics and features that do not make a significant contribution to landscape character or distinctiveness locally or that are untypical or uncharacteristic of the landscape type. Landscapes identified as being generally tolerant of the proposed change subject to design and mitigation, etc.

Source: Tyldesley and Associates 2005.

Such impacts are likely to be considered significant.

Significance is often measured on the yardstick of values (Lawrence 2005), and any process of identifying and evaluating significance must recognize that the determination of significance is inherently an anthropocentric concept (Rossouw 2003). The values and concerns (or judgments) of decision-makers and of the publics underlie interpretations of significance. As Baker and Rapaport (2005) explain, evaluation of significance based strictly on scientific data is inadequate in many cases because resources and ecosystems are linked with human values and cultural meaning. As Hilden (1997) explains, a community directly affected by a proposed undertaking may regard any identifiable effect as highly significant, whereas a community outside of the affected area may have little interest in the proposal. On the other hand, if a development is proposed for a sensitive environment (e.g., active sand dunes or native grasslands), stakeholders who are far removed from the direct effects of development may also consider any identifiable impacts as highly significant. In the latter case, such fears may stem from concern over biodiversity or landscape functioning. In that respect, they reflect the same significance concerns held by decision-makers and experts or reflect the need to consider the sensitivity and value of the receiving environment more closely as the main trigger for assessment. Determination of significance and the need for assessment should include consideration of public concerns and the values they represent; however, in Canadian systems, it is widely held that significance determination must be based on scientific and credible technical information (Sadler 1996).

Desired thresholds. Thresholds need not always be based on standards or regulatory requirements. As already discussed, thresholds for adverse impacts are often stated as *maximum allowable effects levels.* These levels are often based on the desired outcomes and are determined by policy, compatibility with existing plans or programs, or societal objectives. In cases where predicted impacts exceed desired thresholds or maximum allowable change for particular environmental components, the impact is considered significant—even though it may not exceed regulatory standards or thresholds. Haug et al. (1984) proposed a set of criteria to help determine the priority of threshold concern and the probability that an effect will cross the threshold of concern. Each is determined by a maximum or minimum value (quantitative or qualitative). Listed from high to low priority for significance determination, the criteria include:

1. Legal thresholds: established by law or by regulatory limits.
2. Functional thresholds: established for resource use so as to avoid adverse impacts or disturbances that may disrupt ecosystem functioning or destroy resources.
3. Normative thresholds: established by communities or regions based on social norms in relation to environmental, economic, or social concerns.
4. Controversial thresholds: impacts that may have a high profile because of potential conflicts between the affected interest groups but would not otherwise have priority.
5. Individual thresholds: based on the priorities of individuals, groups, or organizations and do not warrant higher priority based on the above reasons.

APPROACHES TO SIGNIFICANCE DETERMINATION

Determining significance is a highly subjective process, yet significance can and should be determined in a consistent and systematic fashion. Several approaches can be used for significance determination, including technical, collaborative, and reasoned argumentation (see Lawrence 2005), and various methods and techniques are available to support such approaches, including Geographic Information Systems, simulation modelling, and statistical significance tests; data scaling and screening procedures, such as threshold analysis and constraint mapping; qualitative and quantitative aggregation and evaluation procedures, including concordance analysis, multi-criteria analysis, ranking and weighting, and risk assessment; and formal and informal public interaction procedures, such as open houses, workshops, and advisory committees (Lawrence 2004). As with many other aspects of EIA practice, there is no specific set of methods and techniques for determining significance. A number of guidelines, however, should be adhered to—namely, that significance thresholds, criteria, and methods be explicit, easy to use, traceable, verifiable, readily understandable, relevant to the problem at hand, and capable of facilitating the interpretation of significance (Sadler 1996).

Technical Approach

Under the technical approach, determining significance typically involves adopting one or more standardized scaled or quantified methods, each of which is based on some characterization of the various dimensions of the anticipated impacts of development, such as impact scale, severity, reversibility, probability, and duration. The technical model captures the most widely used set of standardized methods for significance determination (Whitelaw 1997) and currently forms the basis of ISO 14004 guidance for the evaluation of significance.

Under the technical approach, when a predicted effect meets a pre-established standard or threshold, determining significance is relatively straightforward.

However, in many cases, environmental effects are evaluated only in terms of relative significance. For example, is a short-term effect on human health from factory emissions more significant than a major long-term increase in local employment? It is often the case in EIA that effects and impacts are expressed as 'major' or 'minor' without any evaluation of the relative importance of the environmental components under consideration (Figure 8.2). Yet without some type of weighting approach to determine the **relative significance** of all the environmental components, comparing project action 'i' on environmental components 'x' and 'y' gives little indication of actual significance (Figures 8.3 and 8.4). Some examples of technical approaches to relative significance determination include fixed-point scoring, rating, and paired comparisons. Hajkowicz, McDonald, and Smith (2000) provide a more detailed discussion of weighting methods and techniques.

Fixed-point scoring. In **fixed-point scoring**, a fixed number of values are distributed among all affected environmental components. The higher the point score, the more important the environmental component. In other words, project effects on environmental components assigned high scores are likely to be considered more significant than effects on environmental components with low fixed-point scores. For example, assume four VECs for which a fixed number of points, totalling no more than '1', must be distributed to indicate relative VEC importance:

VEC	Importance (w_i / 1)
water quality	.25
noise	.10
employment	.30
human health	.35

Increasing the importance of any one VEC directly affects the relative significance of project effects on that VEC. At the same time, those assigning the weights, whether EIA decision-makers, experts, or public interest groups, are forced to make trade-offs between VECs because increasing the importance of one VEC requires decreasing the importance of another. While explicit for decision-making purposes, direct trade-offs between VECs may not always be possible.

Rating. With a **rating** approach to assigning impact importance, the importance or significance of each VEC is indicated on a numerical rating scale—for example: 1 = not important; 5 = moderately important; 10 = extremely important; 2, 3, 4, 6, 7, 8, 9 = intermediate values. While rating does not require direct trade-offs in assignment of weights, the disadvantage is that it does not directly indicate the relative importance of one component in comparison to another.

Paired comparisons. As with fixed-point scoring, **paired comparisons** force the decision-maker to consider trade-offs. However, whereas fixed-point scoring requires multiple, simultaneous, and often complex trade-offs for the entire list of environmental components, paired comparisons involve making trade-offs one at a time, for each pair of VECs, thereby contributing to better overall understanding of the decision problem.

The paired comparison approach is based on Saaty's Analytical Hierarchy Process (AHP), a systematic procedure for representing the elements of any problem hierarchically. The AHP organizes the basic rationality by breaking down a problem into

Significance of residual effects

Significance	Symbol
not applicable	blank
negligible	○
negligible/minor	● (heavy ring)
minor	◉
minor/moderate	◎
moderate	◆
moderate/major	✶
major	✹
unknown	⊗

Project Period and Activity — Construction Phase

VECs	blasting	road traffic	roads, dams, infrastructure	quarrying	noise	channel diversion	lake dewatering	pre-stripping	diesel power generation	process plant	human activity	tailings disposal	flow and stream diversion
air quality	negligible	negligible											
permafrost			negligible										
eskers				minor									
water quality			negligible/minor										
fish and aquatic habitat			minor			negligible	negligible/minor						
vegetation			negligible					negligible/minor	negligible				
wildlife and wildlife habitat			negligible/minor							negligible/minor			
caribou			negligible/minor		negligible/minor	negligible/minor				negligible/minor	negligible/minor	negligible/minor	
grizzly bears			minor		minor						negligible/minor	negligible/minor	
wilderness											minor		
biodiversity											negligible		
hydrology						negligible	negligible						minor
climate			negligible				negligible						
groundwater							negligible						

Project Period and Activity — Operations Phase

VECs	process plant operation	diesel power generation	roads and road traffic	tailings water disposal	freshwater supply	waste rock dumping	excavation of pits	winter roads	human activity	diversion channel	noise	flow and stream diversion
air quality	negligible	negligible	negligible									
permafrost				negligible								
eskers												
water quality				negligible								
fish and aquatic habitat			negligible/minor		negligible/minor	negligible						
vegetation		negligible					negligible/minor	minor				
wildlife and wildlife habitat	negligible		negligible									
caribou	negligible		negligible/minor	negligible					negligible/minor	negligible/minor	negligible/minor	
grizzly bears	negligible/minor		minor						minor		minor	
wilderness									minor			
biodiversity									negligible			
hydrology												minor
climate		negligible				unknown						
groundwater												negligible

Figure 8.2 Sample impact significance matrix

Source: Based on BHPB 1998.

Environmental parameter	Project Actions: blasting	site clearing	dredging	road construction	waste disposal	equipment transport	Impact Score
Air quality	–1			–1	–1		–3
Water resources	–2	–3	–3				–8
Water quality	–2	–4	–2				–8
Noise	–2		–1	–2		–2	–7
Forests and vegetation		–5		–3			–8
Wildife	–2	–4		–2			–8
Human health	–2			–1	–4		–7

+ = positive impact
– = adverse impact
no impact =
negligible impact = 1
minor (slight or short-term) = 2
moderate impact = 3
major impact (irreversible or long-term) = 4
severe impact (permanent) = 5

Figure 8.3 Unweighted impact assessment matrix.

Assuming an example of simple additive impacts, the impact of 'site clearing' on 'forest and vegetation' appears to be the most significant individual project impact. In terms of overall project impacts, the impacts of project activities on water resources, water quality, forests and vegetation, and wildlife appear to be equally significant. The relative importance of the affected environmental components have not been determined.

its smaller constituent parts and then guides the decision-maker through a series of pair-wise comparison judgments to express the relative importance of the elements in hierarchy in ratio form, from which decision weights are derived based on the principal eigenvector approach (Saaty 1977). An eigenvector is a linear combination of variables that consolidates the variance, or eigenvalues, in a matrix. The eigenvalues indicate the relative strength (weight) of each of the VECs, where the larger the eigenvalue, the larger the role the paired comparison plays in weighting the entire assessment matrix and therefore determining significance.

For any pair of environmental components, or VECs *i* and *j*, out of the set of components *C*, the individual decision-maker can provide a paired comparison C_{ij} of the components under consideration on a ratio scale that is reciprocal, such that $C_{ji} = 1/C_{ij}$. In other words, if environmental component VEC *i* is 'seven times' more important than VEC *j*, then the reciprocal property must hold—that is, VEC *j* must be seven times less important than VEC *i*. When a decision-maker compares any two VECs *i*, *j*, one VEC can never be judged to be infinitely more important than another. If such a case should arise where VEC *i* is infinitely more than *j*, then no decision tool would be required.

Using the paired comparison approach, decision-makers are presented with a nine-point decision scale ranging from '1', if both components or VECs are equally preferred, '3' for a weak preference of VEC *i* over *j*, '5' for a strong preference, and so forth. If VEC *j* is preferred to *i* for any given criteria, the reciprocal values hold true: 1, 1/3, and 1/5 (Box 8.1). The scale is standardized and unit-free; thus, there is no need to transform all measures, for example, into monetary units for comparative purposes, which is a key advantage of paired comparisons over cost-benefit analysis or utility functions. An additional advantage of paired comparisons over other weighting methods is that the

Environmental Parameter	Weight	Project Actions: blasting	site clearing	dredging	road construction	waste disposal	equipment transport	Impact Score
Air quality	0.16	−1(0.16) = −0.16			−1(0.16) = −0.16	−1(0.16) = −0.16		−0.48
Water resources	0.08	−2(0.08) = −0.16	−3(0.08) = −0.24	−3(0.08) = −0.24				−0.64
Water quality	0.22	−2(0.22) = −0.44	−4(0.22) = −0.88	−2(0.22) = −0.44				-1.76
Noise	0.04	−2(0.04) = −0.08		−1(0.04) = −0.04	−2(0.04) = −0.08		−2(0.04) = −0.08	−0.28
Forests and vegetation	0.08		−5(0.08) = −0.40		−3(0.08) = −0.24			−0.64
Wildife	0.08	−2(0.08) = −0.16	−4(0.08) = −0.32		−2(0.08) = −0.16			−0.64
Human health	0.22	−2(0.22) = −0.44			−1(0.22) = −0.22	−4(0.22) = −0.88		−1.54

+ = positive impact
− = adverse impact

no impact =
negligible impact = 1
minor (slight or short-term) = 2
moderate impact = 3
major impact (irreversible or long-term) = 4
severe impact (permanent) = 5

Figure 8.4 Weighted impact assessment matrix.

Weights are assigned to indicate the importance of the affected environmental components and therefore the relative significance of the impacts. Using the same example as in Figure 8.3, the impact of 'site clearing' on 'forest and vegetation' is no longer the most significant individual project impact. Given the importance of the affected components, the impacts of 'site clearing' on 'water quality' and of 'waste disposal' on 'human health' are now the most significant individual impacts. In terms of overall project impacts, the impacts of project activities on 'water quality' are now the most significant, whereas overall impacts on 'forest resources' and 'wildlife' now appear to be relatively less significant. As a result, the unweighted assessment matrix depicted in Figure 8.3 for determining significance may lead to an incorrect allocation of resources for impact management.

paired comparison approach also provides a measure of consistency or the extent to which weights were assigned purposefully or at random.

While paired comparison generates the most informative weightings for determination of significance, it is also the most complex of the three approaches presented here. In fact, when relying on the public or experts to assign the values, the approach is limited to the number of items that an individual can simultaneously compare—usually 7 +/− 2 (Miller 1956). However, when relying on simulation models or computer-based technology such as Geographic Information Systems, the number of environmental components that can be weighted is limited only by the time and resources available to the EIA practitioner.

Collaborative Approach

The collaborative approach, explains Lawrence (2005, 16), 'starts from the premise that subjective, value-based judgments about what is important should result from

Box 8.1 Paired Comparison Process for Assigning Weights to Environmental Components

The relative importance of component ci in the assessment of project impacts is defined by:
9 = component i is extremely more important than j
7 = component i is very important compared to component j
5 = component i is strongly more important than component j
3 = component i is moderately more important than component j
1 = component i is equally important to component j
1/3 = component j is moderately more important than component i
1/5 = component j is strongly more important than component i
1/7 = component j is very important compared to component i
1/9 = component j is extremely more important than component i

Sample paired comparison assessment matrix for environmental components (VECs)

	VEC A	VEC B	VEC C
VEC A	1	1/3	7
VEC B	3	1	7
VEC C	1/7	1/7	1

For example, VEC B is considered 'moderately more important' than VEC A, so a value of 3 is entered in row 2, column 1; thus 1/3 in row 1, column 2.

To derive the weights for each VEC:

1. Divide each cell entry of the matrix by the sum of its corresponding column to normalize the matrix:

	VEC A	VEC B	VEC C
VEC A	0.24	0.23	0.47
VEC B	0.72	0.68	0.47
VEC C	0.03	0.10	0.07

2. Determine the priority vector of the matrix by averaging the row entries in the normalized assessment matrix:

   Priority vector: A = 0.31; B = 0.62; C = 0.07

3. Multiply each column of the initial paired comparison matrix by its priority and sum the results:

$$0.31 \begin{pmatrix} 1 \\ 3 \\ 1/7 \end{pmatrix} + 0.62 \begin{pmatrix} 1/3 \\ 1 \\ 1/6 \end{pmatrix} + 0.07 \begin{pmatrix} 7 \\ 7 \\ 1 \end{pmatrix} = \begin{pmatrix} 1.01 \\ 2.04 \\ 0.21 \end{pmatrix}$$

4. Divide the results by the original priority vector to determine the relative importance or weight of each VEC:

   VEC A = 1.05/ 0.31 = 3.26
   VEC B = 2.04/ 0.62 = 3.29
   VEC C = 0.21/ 0.07 = 3.00

5. The relative importances are used to interpret the relative significance of project impacts on each of the affected VECs.

interactions among interested and affected parties'. Under this approach, there are no predefined thresholds or criteria for judgment—rather, judgments of what is important, what is acceptable, and what are the limits of allowable change emerge only through consultation with the publics. Issues pertaining to regional and community context are integrated directly into significance determination, and the compatibility of the proposal and its impacts with regard to goals and objectives are central to the decision-making process. As a collaborative approach, the key role of regulators is ensuring the involvement of the most directly affected communities or stakeholders and that all stakeholder concerns are assessed in significance determination.

Numerous methods and approaches are available to facilitate collaborative approaches to significance determination, including open houses and community forums, interactive web-based forums, key informant interviews, community or regional profiling, rapid rural appraisal, nominal group processes, and intervener funding. The overall objectives of using such methods are to fully integrate community, technical, and traditional knowledge in significance determination and to forge a stronger relationship with the public throughout the EIA process. It is important to note, however, that public concerns about environmental effects are not always the same as actual environmental effects resulting from project actions.

Reasoned Argumentation

The reasoned argumentation approach views significance determination as based on reasoned judgments supported by evidence. Lawrence (2005, 19) explains that reasoned argumentation starts from the premise that technical and collaborative models are 'too narrow to provide an adequate foundation for value-based significance judgments about what is important and what is not important'. Usually expressed qualitatively, the reasoned argumentation model is evident at the regulatory level in priorities and objectives of EIA legislation or regulation, often defining 'matters of significance', which are used as triggers during the screening process, and further expressed in project-specific guidelines and requirements. At the practical level, reasoned argumentation, similar to a court decision process, involves sifting through information, data, perspectives, and expressed values using structured methods (e.g., decision support aids, matrices, network diagrams) to focus on matters of most importance to decision-making and to build 'reasoned' arguments that support significance determination—hence the importance of complete and clear documentation of the evaluation process and reasons for the decision (Kontic 2000). The reasoned argumentation model is flexible and responsive to context; however, a well-reasoned argument for significance does not ensure that full consideration has been given to scientific data, public values, or existing technical information.

Composite Approach

The choice of approach to significance determination will vary with context, taking into consideration the nature of the project, the receiving environment, interests, and the political setting. In a study of methods used to determine significance in EIA practices in England, for example, Wood, Glasson, and Becker (2006) found that professional judgment and experience, consultation, and simple checklists of impacts were the methods most commonly used by local planning authorities and consultants. In other words, a composite model consisting of various combinations of technical, collaborative, and argumentative approaches is desirable. Such a model may consist of technical analysis using conventional significance determination methods, supported by public consultation or traditional knowledge systems, which together, based on existing EIA regulation or land-use plans, compile a reasoned argument for significance. Lawrence (2005) suggests that at minimum, effective impact significance determination relies upon:

- use of a variety of technical methods and analytical techniques;
- a range of public consultation methods, including methods that facilitate stakeholder interaction, to support collaborative significance determination;
- use of existing regional, community, or land-use plans or local social surveys to identify values and aspirations against which to compare the proposed development;
- literature analysis and case study reviews of previous, similar proposals and outcomes in comparable environments and situations;
- exploring the uncertainties and acceptable risks associated with the significance determination.

GUIDANCE FOR DETERMINING SIGNIFICANCE

Sadler (1996) identifies several guiding principles for significance interpretation, suggesting, for example, that significance thresholds, criteria, and methods be explicit, easy to use, traceable, substantiated, readily understandable, and relevant to the problem at hand. Methods should be broadly supported and structured and focus the significance interpretation effort, provision should be made to involve interested and affected stakeholders, and the rationale for significance interpretations should be clearly indicated. Building on the work of Sadler and others, Lawrence (2005) provides a more comprehensive list of guiding principles for significance determination:

1. Focused and efficient: Concentrate on aspects most relevant to decision-making and consistent with regulatory and public concern.
2. Explicit: Values and bases for judgments are clear and understandable.
3. Logical and substantiated: Reasoning behind significance determination is logical, and data and analyses are clearly linked to significance judgments.
4. Systematic and traceable: A coherent procedure exists for integrating information and objectives, and that procedure can be reconstructed.
5. Appropriate: Sensitivity of the context is considered in the significance determination.
6. Consistent: Similar projects and issues are treated in a similar manner.
7. Collective and collaborative: Interested and affected parties are involved.
8. Effective: Outcomes of significance determination help to realize public, policy, or EA goals and objectives.
9. Adaptable: Procedures for significance determination are flexible to different contexts and changing circumstances over space and time.

It is not likely that all of these principles can be adhered to simultaneously; however, when determining whether the effects of a proposed development are significant, consideration should be given not only to the characteristics of the effects but also to the contextual factors of public concern, conditions and sensitivity of the receiving environment, specified thresholds and objectives, significance scale, cumulative change and induced effects, contributions to sustainability, and the nature of proposed mitigation measures. Significance is a highly subjective concept, and differences of opinion are likely to continue as to what constitutes a significant environmental effect, how contextual factors are to be considered, what represents an appropriate level of detail for regulatory requirements and standards, and how to make significance determinations that are transparent and robust. The challenges for significance determination do not lie simply in the realms of improved science and the pursuit of objective expert evaluations but in the clarity of communication of the assessment to decision-makers and the stakeholder community as well (Wood 2008). Significance is not so much the search for objectivity as it is how well subjectivity can be substantiated (Lawrence 1993).

KEY TERMS

acid mine drainage
assimilative capacity
fixed-point scoring
impact significance
paired comparisons
rating
relative significance
residual effects
statistical significance

STUDY QUESTIONS AND EXERCISES

1. Determining the significance of environmental effects is often a subjective process, particularly when issues of social concern are involved. Suppose, for example, that a proposed development is likely to result in the closure of a local outdoor public recreational area. How might you assess the significance of such an impact?
2. Obtain a completed project EIS from your local library or government registry, or access one on-line. Identify the types of criteria used to determine impact significance. Compare your findings to those of others. Are there noticeable similarities of significance criteria across impact statements? Is there evidence of 'sustainability criteria'?
3. Impact significance in EIA is often classified as 'major', 'moderate', 'minor', or 'negligible', without any qualification as to what these terms mean. Following the example illustrated in Figure 8.2, construct a simple impact significance matrix for the construction and operation of a waste incineration project. Identify project actions across the top and VECs down the side. Develop a legend for impact significance similar to that in Figure 8.2, and provide operational definitions for each level of significance based on the significance criteria discussed in this chapter.
4. Discuss the advantages of weighted impact assessment matrices over unweighted impact assessment matrices.
5. Construct a 3 × 2 table to compare and contrast the relative advantages and disadvantages of technical, collaborative, and reasoned argumentation approaches to significance determination.

6. Assume that the following VECs have been identified for a large-scale energy project to be developed in your region:
   - air quality
   - employment
   - forests and vegetation
   - water quality
   - human health

   a) Use fixed-point scoring to assign weights to the affected VECs so that the total of all weights equals 1.
   b) Use a numerical rating scale to assign weights to the affected VECs so that 1 = not important, 3 = moderately important, 5 = important, 7 = very important, and 9 = extremely important.
   c) Construct a paired comparison matrix, and using the scale depicted in Box 8.1, calculate weights for each of the VECs.
   d) For each of the above approaches, standardize each of the weights for each VEC by using the following scaling parameter: $(i - i_{min}) / (i_{max} - i_{min})$, where i is the respective weight and i_{min} and i_{max} represent the minimum and maximum values of all weights, respectively, for the set of VECs. This will generate a standardized scale in which the least important VEC = 0 and the most important VEC = 1.
   e) Compare your results across weighting techniques. Are the weights different? Why?
   f) Discuss the advantage of using paired comparisons over the other two weighting techniques.
7. Who should be responsible for determining impact significance? Are there certain criteria that would apply to practically all proposed developments in your area? What is the role of the public in determining impact significance? Think back to the principles of public involvement discussed in Chapter 4.

REFERENCES

Baker, D., and E. Rapaport. 2005. 'The science of assessment: Identifying and predicting environmental impacts'. In K. Hanna, Ed., *Environmental Impact Assessment Practice and Participation*. Toronto: Oxford University Press.

Benson, W. 2004. 'Determining significant effects within environmental management: A case study of the UK Ministry of Defense'. Proceedings of the International Sustainable Development Research Conference, University of Manchester.

Byron, H. 2000. *Biodiversity and Environmental Impact Assessment: A Good Practice Guide for Road Schemes*. Sandy, UK: The RSPB, WWF-UK, and English Nature and Wildlife Trusts.

Canter, L.W. 1996. *Environmental Impact Assessment*. 2nd edn. New York: McGraw-Hill.

Canter, L.W., and G.A. Canty. 1993. 'Impact significance determination—Basic considerations and a sequenced approach'. *Environmental Impact Assessment Review* 13: 275–97.

Deakin, M., S. Curwell, and P. Lombari. 2002. 'Sustainable urban development: The framework and directory of assessment methods'. *Journal of Environmental Assessment Policy and Management* 11 (3): 171–97.

Duinker, P.N., and G.E. Beanlands. 1986. 'The significance of environmental impacts: An exploration of the concept'. *Environmental Management* 10 (1): 1–10.

FEARO (Federal Environmental Assessment Review Office). 1994. 'Reference guide: Determining whether a project is likely to cause significant adverse environmental effects'. In Canadian Environmental Assessment Agency, Ed., *The Canadian*

Environmental Assessment Act Responsible Authority's Guide. Ottawa: Ministry of Supply and Services Canada.

Gibson, R.B. 2001. *Specification of Sustainability-Based Environmental Assessment Decision Criteria and Implications for Determining 'Significance' in Environmental Assessment*. A report prepared under a contribution agreement with the Canadian Environmental Assessment Agency Research and Development Program. Gatineau, QC: CEAA.

Hajkowicz, S.A., G.T. McDonald, and P.N. Smith. 2000. 'An evaluation of multiple objective decision support weighting techniques in natural resource management'. *Journal of Environmental Planning and Management* 43 (4): 505–18.

Haug, P.T., et al. 1984. 'Determining the significance of environmental issues under the National Environmental Policy Act'. *Journal of Environmental Management* 18: 15–24.

Hilden, M. 1997. 'Evaluation of the significance of environmental impacts'. Report of the EIA process strengthening workshop, 4–7 April, Canberra, Australia.

Irwin, F., and B. Rodes. 1992. *Making Decisions on Cumulative Environmental Impacts: A Conceptual Framework*. Washington: World Wildlife Fund.

Kjellerup, U. 1999. 'Significance determination: A rational reconstruction of decisions'. *Environmental Impact Assessment Review* 19: 3–19.

Kontic, B. 2000. 'Why are some experts more credible than others?' *Environmental Impact Assessment Review* 20: 427–34.

Lawrence, D.P. 1993. 'Quantitative versus qualitative evaluation: A false dichotomy?' *Environmental Impact Assessment Review* 13 (1): 3–12.

———. 2000. 'Significance in environmental assessment'. Research and Development Monograph Series. Gatineau, QC: Canadian Environmental Assessment Agency.

———. 2004. *Significance in Environmental Assessment*. Research supported by the Canadian Environmental Assessment Agency's Research and Development Program for the Research and Development Monograph Series, 2000. Gatineau, QC.: CEAA.

———. 2005. 'Significance criteria and determination in sustainability-based environmental impact assessment'. Report to the Mackenzie Gas Project Joint Review Panel. Langley, BC: Lawrence Environmental.

Miller, G.A. 1956. 'The magical number seven plus or minus two: Some limits on our capacity for processing information'. *Psychological Review* 63: 81–97.

Morrison-Saunders, A., et al. 2001. 'Roles and stakes in environmental impact assessment follow-up'. *Impact Assessment and Project Appraisal* 19 (4): 289–96.

Ross, W.A., A. Morrison-Saunders, and R. Marshall. 2006. 'Common sense in environmental impact assessment: It is not as common as it should be'. *Impact Assessment and Project Appraisal* 24 (1): 3–22.

Rossouw, N. 2003. 'A review of methods and general criteria for determining impact significance'. *AJEAM-RAGEE* 6: 44–61.

Saaty, T.L. 1977. 'A scaling method for priorities in hierarchical structures'. *Journal of Mathematical Psychology* 15: 243–81.

Sadler, B. 1996. *Environmental Assessment in a Changing World: Evaluating Practice to Improve Performance*. Final report of the International Study of the Effectiveness of Environmental Assessment. Fargo, ND: IAIA.

Sippe, R. 1999. 'Criteria and standards for assessing significant impact'. In J. Petts, Ed., *Handbook of Environmental Impact Assessment*, v. 1, *Environmental Impact Assessment: Process, Methods and Potential*. London: Blackwell Science.

Tinker, L., et al. 'Impact mitigation in environmental assessment: Paper promises or the basis of consent conditions?' *Impact Assessment and Project Appraisal* 23 (4): 265–80.

Tyldesley and Associates. 2005. *Environmental Assessment Handbook: Guidance on the Environmental Assessment Process*. Doc Ref: 1477 4th edition, Issue: 04A.

US Council on Environmental Quality. 1973. *Preparation of Environmental Impact Statements: Guidelines*. Washington: Council on Environmental Quality.

———. 1987. 'Regulations for implementing NEPA'. Section 1508.27.40 Code of Federal Regulations, 1987. http://ceq.eh.doe.gov/nepa/regs/ceq/toc_ceq.htm.

Voisey's Bay Nickel Company. 1997. *Voisey's Bay Mine/Mill Project Environmental Impact Statement*. St John's: Voisey's Bay Nickel Company.

Westman, W.E. 1985. *Ecology, Impact Assessment and Environmental Planning*. New York: John Wiley and Sons.

Whitelaw, K. 1997. *ISO 14001 Environmental System Handbook*. Oxford: Butterworth-Heinnemann.

Wood, C. 1995. *Environmental Impact Assessment: A Comparative Review*. London: Longman Scientific and Technical.

Wood, C., and J. Becker. 2004. 'Evaluating and communicating impact significance in EIA: A fuzzy set approach to articulating stakeholder perspectives'. Paper presented at the annual meeting of the International Association for Impact Assessment, 26–29 April.

Wood, G. 2008. 'Thresholds and criteria for evaluating and communicating impact significance in environmental statements: See no evil, hear no evil, speak no evil'. *Environmental Impact Assessment Review* 28: 22–38.

Wood, G., J. Glasson, and J. Becker. 2006. 'EIA scoping in England and Wales: Practitioner approaches, perspectives and constraints'. *Environmental Impact Assessment Review* 26: 221–41.

CHAPTER 9

Managing Project Impacts

IMPACT MANAGEMENT

Once potential environmental effects and impacts have been identified and their significance determined, the next phase of the EIA process is to design management strategies to address those effects and impacts. Impact management is foundational to the entire EIA process in that it requires the identification of impact management measures that translates the findings from an EIS into recommendations to avoid, minimize, or offset those impacts (Tinker et al. 2005). Impact management involves plans or strategies designed to avoid or alleviate anticipated impacts generally perceived as undesirable and to generate or enhance effects seen as beneficial (Box 9.1).

The determination of potential effects and impacts identified through the environmental assessment process is often imprecise because of uncertainties surrounding such factors as the project's design and timetable or because of exogenous factors. Thus, a management system designed to address central issues should be flexible enough to respond to unanticipated impacts as well as differences between the actual and predicted nature, level, or significance of the impacts.

Box 9.1 Potential Impacts and Management Approaches for a Road Construction Project

Component/Action	Impact	Management
dust	residential air quality worker health	water gravel surfaces require safety masks
habitat	vegetation destruction risk to wetlands	restore or create elsewhere avoid by rerouting
wildlife	habitat fragmentation vehicle collisions	create wildlife overpasses road signage
equipment hours	noise disruption emissions worker safety	schedule construction vehicle standards inspections and training
gravel infilling	stream sedimentation	sediment screens
economics	employment	hire locally

MANAGING ADVERSE IMPACTS

Avoidance

Avoiding potentially adverse environmental effects and impacts at the outset is clearly the most desirable approach—if an impact can be avoided, then the time and financial resources required by impact mitigation or compensation can also be avoided. Methods of avoiding potentially adverse impacts can include such measures as setting regulatory standards concerning the use of toxic substances, scheduling project construction activities so that they do not conflict with daily patterns of local socio-economic activity, land-use planning and zone designation, and the construction of self-contained work camps to avoid potentially negative socio-economic effects that might be caused by site worker–community interaction. **Impact avoidance** does not necessarily enter the equation only after potential impacts are predicted and their significances determined, since much avoidance can occur early in the scoping process through identifying alternative locations, project designs, or implementation strategies. For example:

Activity: road construction
Causal factor: surface disturbance
Effect/impact: loss of wetland habitat and function
Avoidance: re-routing to avoid wetland habitat

Mitigation

Not all potentially adverse impacts are recognized in advance. While mitigation is often used in a generic way to refer to impact management in general, strictly speaking, **impact mitigation** is defined as the application of project design (Box 9.2), construction, operation, scheduling, and management principles and practices to *minimize* potentially adverse environmental impacts. For example, forest-harvesting operations can lead to soil erosion and excessive runoff, which in turn may affect the quality of aquatic environments. The maintenance of **buffer zones**, or areas of undisturbed vegetation, is thus a desired impact mitigation practice in forestry. While buffer zones do not *prevent* soil erosion or surface runoff, they do reduce the severity of the impacts of erosion and runoff on aquatic environments. For example:

Activity: forest clearing
Causal factor: erosion and runoff
Effect/impact: increased turbidity in local stream
Mitigation: stream buffer zones

Rectification

Not all adverse environmental effects can be avoided or mitigated. In certain cases, environmental components will inevitably be temporarily damaged. **Rectifying impacts** refers to restoring environmental quality, rehabilitating certain environmental features, or restoring environmental components to varying degrees. For example, in cases where the construction phase of project development requires

Box 9.2 Mitigating Visual Impacts through Project Design

Declared a World Heritage Site by UNESCO in 1981, the Head-Smashed-In Buffalo Jump at Waterton Park, Alberta, is one of the largest and best preserved buffalo jumps in North America. Construction of an interpretative centre commenced on the site in 1986 and concluded near the end of that year. A major concern for building architects and landscape planners was minimizing the visual impact of the interpretation centre on the bare, rolling plains of the prairie landscape. Consequently, the building was designed to represent a harmonious relationship with the surrounding landscape. The building itself is totally submerged into the site, blending into the landscape both in design and colour. Visitors approach the main entrance of the building at the foot of a small cliff and proceed through an entrance flanked on both sides by retaining walls up to 10 metres high simulating bedrock scars depicting an archaeological dig site.

clearing of forested landscape, impact management efforts can focus on restoring the landscape post-construction to resemble the pre-disturbed state. In the case of a **sanitary landfill** site, for example, the site is typically returned to a vegetative state once the landfill is full and covered. While the rectified site condition may not exactly resemble the pre-disturbed condition, the objective is to return it to a more desirable condition than what was created by the project actions. For example:

Activity: landfill operation
Causal factor: site preparation and waste disposal
Effect/impact: habitat loss
Rectification: site reclamation with native vegetation following decommissioning

Compensation

Some environmental effects cannot be avoided, mitigated, or rectified. In such cases, the typical action is **compensation** for the unavoidable, residual, or irreparable impacts that remain after other impact management options have been exhausted or for which no management alternative exists. Compensation sometimes involves monetary or other benefit payments to those affected by the damage caused by the project, while in other cases it involves measures to recreate environmental habitats at an alternative site. In Canada, for example, a federal policy on aquatic-based habitat declares that there should be 'no net loss'. This is not to say that projects posing a threat to the destruction of aquatic habitat will not be approved; rather, any habitat that is lost must be compensated for. For example:

Activity: open-pit mine development
Causal factor: drainage of a nearby lake for tailings disposal
Effect/impact: loss of fish habitat and aquatic life
Compensation: creation of new habitat, or restoration of degraded habitat, elsewhere in the project region

MANAGING POSITIVE IMPACTS

Creating Benefits

Impact management provides the necessary means not only to reduce or avoid potentially negative impacts but also to create positive benefits from project development. The most desirable approach to managing positive impacts is to ensure that the project makes a positive contribution to the environment and society through the creation of new benefits, such as community economic growth or environmental improvement. For example:

Activity: hydroelectric construction project
Causal factor: materials and services purchasing
Effect/impact: increased local expenditures
Benefit creation: small business development support centre

Enhancing Benefits

Development projects often create as many positive impacts, particularly economic ones, as they do negative impacts. Indeed, if this were not the case, democratic societies and their governments would be hard-pressed to justify the continued approval of such projects. Thus, an important management strategy is to enhance the benefits of potentially positive impacts and to maximize the duration of those impacts. This might involve, for example, ensuring the greatest possible distribution of financial benefits among affected communities over the longest period of time. This was the case for the Voisey's Bay project, discussed in Chapter 1, which for the first time adopted an explicit sustainability mandate requiring the project proponent to enhance potentially positive impacts by committing to community infrastructure investment and worker training programs beyond the life of the mine project. Other common approaches include adopting 'buy local' and 'hire local' strategies. This was an important part of enhancing the benefits of the Hibernia offshore oil platform construction project at Bull Arm, Newfoundland, in which Mobil Oil committed to ensuring that a minimum number of construction contracts would be let to local businesses and that a certain percentage of workers employed at the site would be Newfoundlanders. For the Ekati diamond mine in the Northwest Territories, similar impact enhancement strategies were adopted to maximize the benefits of project employment. For example:

Activity: diamond mine development and operation
Causal factor: job creation
Effect/impact: increased local employment and income
Benefit enhancement: provision of free financial counselling to employees

MEASURES THAT SUPPORT IMPACT MANAGEMENT

In addition to project-based impact management strategies, a number of broader measures facilitate management of the environmental impacts of industry operations.

Perhaps the most internationally recognized measure is an environmental management system.

Environmental Management Systems

The International Organization for Standardization, commonly referred to as ISO, is a global federation of more than 100 countries and is headquartered in Geneva, Switzerland. The organization was formed in 1947 to promote the development of international standards, primarily in product manufacturing. The first system for management standardization (ISO 9000 series) was introduced in the early 1990s. By the mid-1990s, the ISO 14000 series was introduced to promote international industry standards in environmental management, and in 1996 the **ISO 14001** standardization for **environmental management system** (EMS) certification was approved (Box 9.3). Currently, ISO 14001 specifies the requirements and standards for an EMS, and ISO 14004 provides the guidelines for EMS implementation. To date, more than 130,000 ISO 14001 certificates have been issued in 140 countries, with more than 23,000 certificates issued in Japan, followed by China with nearly 20,000. As of 2006, the United States ranked seventh globally, with more than 5,500 certificates, and 1,700 certificates had been issued in Canada (International Organization for Standardization 2007).

Box 9.3 Environmental Management Systems

An EMS is ideally a cyclical process of continual improvement in which an organization is constantly reviewing and revising its management system. This consists of a 'plan-do-check-act' cycle in which the first step is developing a management policy. Through policy development, the organization's environmental goals and objectives are formulated, reflecting the important environmental aspects and targets as well as the legal requirements of the organization. This is followed by planning, implementation, monitoring, and program review. In this sense, not only can an EMS act as a regulatory system by which an organization seeks to meet industry standards, it can also be a potentially valuable tool for environmental management and improvement of industry operations.

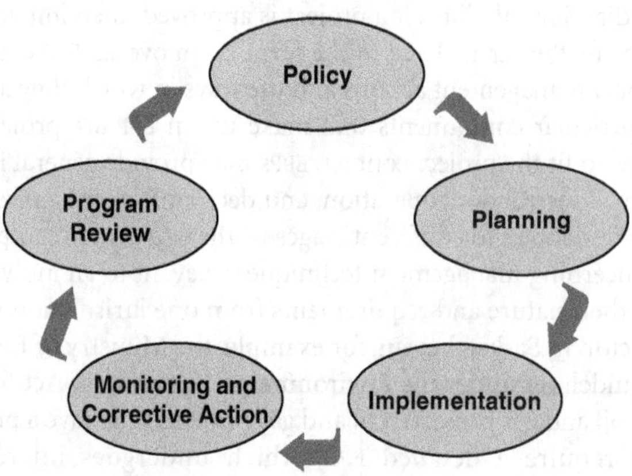

EMSs are currently among the most widely recognized tools for managing the environmental affairs of industry. An EMS is a voluntary industry-based management system by which an organization can:

- identify and control the environmental impact of its operations and activities;
- continually improve its environmental performance;
- implement a systematic plan to set environmental targets and objectives and to clearly demonstrate to the public that those targets and objectives have been achieved.

By putting an EMS into effect, a company is seeking to minimize the environmental impacts of its operations. EMSs emerged in response to industry's realization of the need for an integrated and proactive approach to managing industry environmental issues to assist in ensuring that it is complying with environmental regulations, adopting and following environmental objectives, and receiving economic benefits from improved environmental performance (Strachan, McKay, and Lal 2003). A well-designed EMS should help an organization to develop a proactive approach to environmental management, achieve a balanced view across all aspects of operations, enable effective and directed environmental goal-setting, and contribute to a more effective environmental auditing process. That said, the link between meeting ISO standards and genuine improvement in environmental performance has not been clearly established. A study of toxic emissions from US automobile assembly facilities revealed that facilities with ISO 14001 EMS certification often fared worse than those without certification (Matthews, Christini, and Hendrickson 2004).

Environmental Protection Plans

A second measure to ensure impact management effectiveness is an **environmental protection plan** (EPP). While an EMS is a voluntary industry initiative, an EPP is often a mandatory requirement in project-based EIA. As part of the EIA process, key impacts, issues, and management measures to address those impacts will have been identified. In certain Canadian jurisdictions, if a project is approved, then impact management measures may be further articulated in the form of an overall EPP designed to detail the specific impact management actions and the ways in which they are to be implemented. The particular components that make up an EPP are project-specific and custom-designed to fit the project context. EPPs may provide general information on the management, construction, operation, and decommissioning of certain types of project, may be developed for different stages of the project, or can provide specific information concerning management techniques relevant to an individual project.

EPPs vary in their nature and requirements from one jurisdiction to the next and by industrial sector. In Saskatchewan, for example, the Ministry of Environment has implemented guidelines under the Environmental Assessment Act for the preparation of EPPs for oil and gas projects. Oil and gas projects that have a potential to trigger a full EIA require a detailed EPP, which undergoes interagency review co-ordinated by the Environmental Assessment Branch. The EPP is prepared by the proponent and describes how the project will be undertaken, including such aspects as a description of the project environment, potential conflicts, and environmental protection measures that the proponent will take to avoid or minimize those conflicts. Saskatchewan's 'Guidelines for preparation of an environmental protection plan for oil and gas activities' can be accessed on the province's publications website at http://www.publications.gov.sk.ca/index.cfm.

Impact Benefit Agreements

A third approach to facilitating impact management is an **impact benefit agreement** (IBA), sometimes referred to as a socio-economic agreement, environmental agreement, or impact management agreement. IBAs are particularly useful for addressing the impacts generated by development projects on local communities and are usually associated with compensation measures. IBAs are legally binding agreements between a proponent and a community that serve to ensure that communities have the capacity and resources required to maximize the potential positive benefits stemming from project development.

Negotiated IBAs and environmental agreements are now almost commonplace in the mining sectors of Canada, Australia, and the US and are beginning to emerge in developing nations and in other industry sectors (O'Faircheallaigh and Corbett 2005). In the Canadian context, five agreements for major projects have been implemented during the past decade—for the Ekati, Diavik, and Snap Lake diamond mines in the Northwest Territories, the Voisey's Bay nickel mine in Newfoundland and Labrador, and the Horizon Oil Sands Project in Alberta. All five agreements establish goals and mandates that relate to Aboriginal participation in environmental management, follow-up, and adaptive management; only two (Diavik and Snap Lake) included government as a signatory; none emerged because of a legal requirement for the negotiation of an agreement. Arguably, the rise in such agreements is a reflection of the privatization of environmental governance.

Some of the first IBAs in Canada, such as the one between Ontario Hydro and the Township of Atikokan, Ontario, in 1978, focused primarily on local employment opportunities and community investment. More recent agreements, however, also include environmental restrictions, social and cultural programs, dispute resolution mechanisms, and provisions for revenue-sharing with others (Sosa and Keenan 2001) (Box 9.4). In Canada, IBAs are especially important with regard to industry obtaining the cooperation of Aboriginal communities and for these communities to achieve certain guaranteed benefits. As Veiga, Scoble, and McAllister (2001, 191) explain, 'mining companies must now pursue their interests in a way that also promotes those of the local communities in regions where they are operating.' They go on to suggest that '. . . the challenge for any mining company is to engage in an equitable partnership with the associated community and thus leave a lasting legacy of sustainability and well-being to the community, avoiding environmental degradation and social dislocation.'

The rise in IBAs in recent years may be attributed in part to the deficiencies of negotiating community issues, impacts, and benefits at the time of the development proposal and impact assessment. In a review of impact assessment and benefit agreements in the Mackenzie Valley of the Northwest Territories, for example, Galbraith, Bradshaw, and Rutherford (2007) found that IBAs addressed many issues of concern

Box 9.4 Impact Management Agreement in the Athabasca Uranium Industry

The Athabasca Basin of northern Saskatchewan has been the focus of the uranium mining industry since pitchblende was discovered on the northern shore of Lake Athabasca in the 1930s. The Athabasca region is now one of the world's most important sources of uranium, accounting for approximately 30 per cent of global production.

Saskatchewan's uranium mines are primarily owned and operated by Cameco Corporation and Areva Resources Canada and include McLean Lake, Key Lake/McArthur River, and Rabbit Lake operations. In 2007, a total of 9,465 tonnes of uranium (24.6 million pounds of U_3O_8) was produced at these mines.

In 1993, emerging from recommendations of a joint federal–provincial panel on uranium mining in northern Saskatchewan, the Athabasca Working Group (AWG) was formed to bring together the communities of northern Saskatchewan and the uranium mining industry in discussions about uranium mining activities and to manage the impacts of uranium operations on the environment and communities of the Athabasca region. The AWG represents communities of the Athabasca Basin: Black Lake Denesuline Nation, Fond du Lac Denesuline, and Hatchet Lake Denesuline, along with the northern settlements of Camsell Portage, Uranium City, and Wollastin Lake and the northern hamlet of Stony Rapids.

In 1999, a comprehensive impact management agreement was signed between the communities and the uranium industry. The agreement focuses on three major areas: environmental protection; employment, training, and business development; and benefits-sharing. Benefits-sharing includes education, training, health, culture, recreation, and economic development opportunities. According to Parsons and Barsi (2000), the agreement addressed many socio-economic and community issues in the Athabasca region that the provincial government has failed to address directly—specifically, supporting communities in developing skills and resources so that northerners could receive benefits from mining activities. In this regard, the agreement went well beyond any EIA process and any legal requirements for mine development or related environmental, socio-economic, or mineral regulations.

Benefit-sharing	**Item**
Education	Student employment, scholarships, cultural camps
Training	Skills training, work placements, special apprenticeships, supervisory development
Health	Community and family health, fundraising and donations for health care facilities
Culture	Cultural events, language retention, Elder counsellors
Recreation	Facility development, recreation and sport development, regional sport and recreational events
Economic development	Regional business study

Sources: AWG 2004; Birk 2009; Cameco Corporation 2007; Parsons and Barsi 2000; SMA 2007.

to the local community, including community–industry relationships and benefits-sharing, that project EIAs did not.

IBAs are not meant to replace project EIA but to complement the existing regulatory process. These agreements are negotiated for different reasons depending on the particular community's interests or First Nation's land and resource rights, the regulatory framework in place, and the relationship that exists between the community and the company (Sosa and Keenan 2001). A major challenge to the impact assessment community in understanding the value added by these agreements to the EIA process is that IBAs and similar negotiated agreements occur outside the public realm and little is known about their content, benefits-sharing and impact management details, and overall efficacy.

EIA AND IMPACT MANAGEMENT IN CANADA

EIA is more than just a tool to identify and predict potential environmental impacts; it should also play an important environmental management role. Impact management must be planned in an integrated and coherent fashion to ensure that such measures are effective and non-contradictory and that they do not shift the problem from one environmental component or sector of society to another (Glasson, Therivel, and Chadwick 1999). Thus, impact management is not limited to any one point in the EIA process. From the outset of project design and scoping, measures should be considered for avoiding potentially adverse impacts and for creating or enhancing positive ones.

In practice, the focus of impact management tends to be on reducing or making less severe the adverse environmental effects of development. Storey and Noble (2004) note that while most biophysical effects arising from human activities are adverse, many social, economic, and other human effects are not. Increased employment, training, and business development, for example, are usually regarded as positive outcomes. Mitigation, strictly defined, means to 'make less severe'. While mitigation and making impacts 'less severe' are important, it is only one of several types of action that those responsible for managing project effects may need or wish to take. The emphasis on simply reducing the severity of effects reflects a 'damage control' perspective rather than a proactive one in which actions taken might conceivably have a positive effect. If sustainable development is truly a goal of EIA, then there is a need to adopt a broader perspective on management strategies for all types of potential effects.

Rather than simply aiming to mitigate adverse effects, managers should consider a full range of management options, and the language of EIA should reflect this. As a guide to good practice, Mitchell (1997) suggests a hierarchy of impact management practices, notably:

- avoiding impacts at the source;
- reducing impacts at the source;
- mitigating impacts on-site;
- mitigating impacts on the environmental receptor;
- repairing impacts;
- compensating for impacts;
- enhancing positive impacts.

From a broader sustainability perspective, it might be argued that at the top of this hierarchy should be the creation and enhancement of overall, long-term positive benefits to the environment and society and recognition of the trade-offs involved in doing so.

That said, impact management measures are of little value unless they are actually implemented, monitored for effectiveness, and adjusted accordingly. In practice, impact management is often a condition of development approval, relying on the goodwill of the developer (see Tinker 2003), rather than an obligation that must be met following project implementation.

KEY TERMS

buffer zone
compensation
environmental management system
environmental protection plans
impact avoidance
impact benefit agreement
impact mitigation
ISO 14001
rectifying impacts
sanitary landfill

STUDY QUESTIONS AND EXERCISES

1. Suppose you are responsible for negotiating an IBA for a small community (population less than 5,000) about to be the recipient of a large mining operation. What items might you want to negotiate with the proponent and the government for inclusion in the IBA?
2. Assume a simple highway construction project through a small community. Develop a list of project activities and potential impacts, and propose as many different types of impact management measures as possible, from avoidance to enhancement.
3. Obtain a completed project EIS from your local library or government registry, or access one on-line. Document the nature of impact management measures. For example, are most impact management measures based on avoidance, mitigation, rectification, or compensation? Are there any impact management measures that emphasize creating or enhancing positive project impacts? Generate a list and compare your results with those of others.
4. Visit the International Organization for Standardization website at www.iso.org.
   a) How many countries were ISO members this past year?
   b) Is your country an ISO member?
   c) What is the trend in ISO 14001 certification over the past five years?
   d) What does ISO identify as the principal benefits of ISO 14001 certification to businesses?

REFERENCES

AWG (Athabasca Working Group). 2004. 'Athabasca Working Group annual report'. www.cameco.com.

Birk, J. 2009. 'Role of environmental agreements in environmental impact assessment follow-up: Case study of Saskatchewan's uranium mining industry'. MSc thesis, Department of Geography and Planning, University of Saskatchewan, Saskatoon.

Cameco Corporation. 2007. 'Community dialogue: Athabasca Working Group'. http://www.cameco.com.

Galbraith, L., B. Bradshaw, and M. Rutherford. 2007. 'Towards a supraregulatory approach to environmental assessment in northern Canada'. *Impact Assessment and Project Appraisal* 25 (1): 27–41.

Glasson, J., R. Therivel, and A. Chadwick. 1999. *Introduction to Environmental Impact Assessment: Principles and Procedures, Process, Practice and Prospects*. 2nd edn. London: University College London Press.

International Organization for Standardization. 2007. 'The ISO survey of certification: 2006'. http://www.iso.org/iso/store.htm, under 'products and services/additional publications/management standards'.

Matthews, D.H., G.C. Christini, and C.T. Hendrickson. 2004. 'Five elements for organizational decision-making with an environmental management system'. *Environmental Science and Technology* 38 (7): 1,927–32.

Mitchell, B. 1997. *Resource and Environmental Management*. Harlow, UK: Addison Wesley Longman.

O'Faircheallaigh, C., and T. Corbett. 2005. 'Indigenous participation in environmental management of mining projects: The role of negotiated agreements'. *Environmental Politics* 14 (5): 629–47.

Parsons, G., and R. Barsi. 2000. *Mining and the Community: Uranium Mining in Northern Saskatchewan*. Regina: Organization for Western Economic Cooperation.

SMA (Saskatchewan Mining Association). 2007. 'Commodities mined in Saskatchewan—Uranium reserves as of December 31, 2007'. http://www.saskmining.ca/commodities.php.

Sosa, I., and K. Keenan. 2001. 'Impact benefit agreements between Aboriginal communities and mining companies: Their use in Canada'. http://www.cela.ca/international/IBAeng.pdf.

Storey, K., and B. Noble. 2004. *Toward Increasing the Utility of Follow-up in Canadian EA: A Review of Concepts, Requirements and Experience*. Report prepared for the Canadian Environmental Assessment Agency. Gatineau, QC: CEAA.

Strachan, P.A., I. McKay, and D. Lal. 2003. 'Managing ISO 14001 implementation in the United Kingdom Continental Shelf (UKCS)'. *Corporate Social Responsibility and Environmental Management* 10: 50–63.

Tinker, L. 2003. 'Mitigation is the heart of the environmental impact assessment process, but is the English system effective at ensuring that mitigation measures are put in place? An assessment, using EISs from five countries'. MSc thesis, University of East Anglia, Norwich, UK.

Tinker, L., et al. 2005. 'Impact mitigation in environmental assessment: Paper promises or the basis of consent conditions?' *Impact Assessment and Project Appraisal* 23 (4): 265–80.

Veiga, M., M. Scoble, and M. McAllister. 2001. 'Mining with communities'. *Natural Resources Forum* 25: 191–202.

CHAPTER 10

Post-Decision Monitoring

FOLLOW-UP

Much of what has been covered in this book has focused on what happens before a decision is made to approve or not to approve a project. This has typically been the focus of EIA—a 'build it and forget about it' syndrome. In this regard, EIA is no more than a linear, predictive process with no mechanism to ensure quality assurance of the process itself, verify impact predictions, or evaluate the effectiveness of measures proposed to manage actual project impacts. Only in recent years has attention been given to the follow-up stage of EIA—what happens after a decision is made—but there are still only limited requirements regarding post-decision analysis.

In essence, follow-up is the element that can transform EIA from a static to a dynamic process and the missing link between EIA and effective **life-cycle assessment** (Arts, Caldwell, and Morrison-Saunders 2001). The rationale for follow-up is similar to that of EIA itself—trying to come to grips with the uncertainties intrinsic to a particular activity (Arts, Caldwell, and Morrison-Saunders 2001). Some EIA practitioners interpret follow-up as strictly ensuring that mitigation measures identified in the assessment have been implemented, while others view follow-up as an umbrella that covers the activities, such as routine monitoring or auditing, undertaken during the post-decision stages of the environmental assessment process.

In Canada under the Canadian Environmental Assessment Act, a follow-up program means a program for (1) verifying the accuracy of the environmental assessment of a project and (2) determining the effectiveness of any measures taken to mitigate the adverse environmental effects of the project. Regarding the first point, one question that emerges is whether there is much value in verifying the accuracy of impact predictions when project environmental conditions are constantly changing. Thus, it is perhaps the second aspect of follow-up, essentially a management function, that is of greater value. This is the perspective adopted in this chapter, which views follow-up as largely an adaptive process of mitigation performance evaluation, state-of-the-art environment monitoring, and ongoing revision of mitigation programs and project impact management measures (Figure 10.1).

The Canadian Environmental Assessment Agency's updated Operational Policy Statement (2007) for follow-up programs under the Canadian Environmental Assessment Act adopts a broader perspective than the Act itself, stating that the purposes of a follow-up program are to:

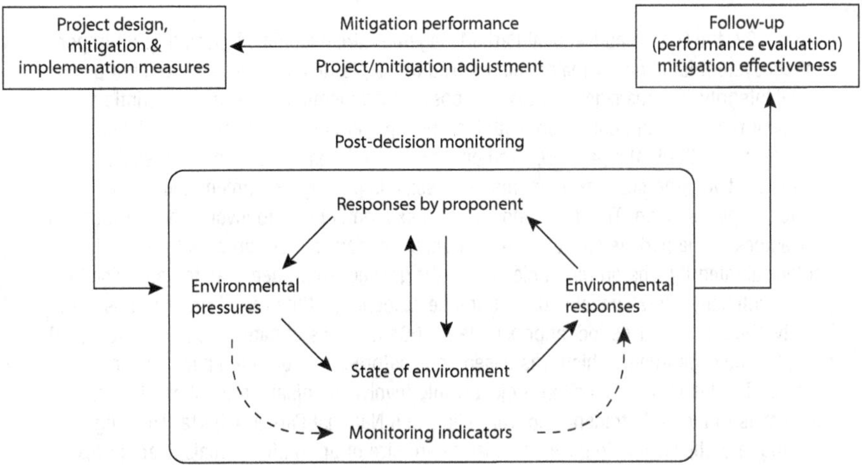

Figure 10.1 Follow-up and monitoring in EIA.

Source: Based on Ramos, Caeiro, and de Melo 2004.

- verify the predictions made in an environmental assessment;
- determine the effectiveness of mitigation measures in order to modify or implement new ones if required;
- support the implementation of adaptive management measures;
- provide information on environmental effects and mitigation measures that can be used to support future assessments;
- support project environmental management systems.

Follow-up programs are mandatory under the Canadian Environmental Assessment Act for all projects assessed by a comprehensive study, mediation, or review panel. For screening assessments, however, the need for a follow-up program is determined on a case-by-case basis by the responsible authority (Box 10.1).

Box 10.1 Following up on Bison Reintroduction and Experimental Grazing in Grasslands National Park, Saskatchewan

Grasslands National Park is located in southwest Saskatchewan, near the Saskatchewan–Montana border in the Great Plains grassland biome. The park covers approximately 906 square kilometres of mixed grass prairie and is divided into two separate blocks. The park is a semi-arid ecosystem that has evolved over time with migratory and sedentary bison grazing, a disturbance considered to have contributed to the heterogeneity and ecological integrity of the landscape. However, bison have been extirpated from the grasslands area since the turn of the nineteenth century, and most of the region outside the park has been cultivated for crop production. The remaining native vegetation has been subjected to grazing by domestic animals.

continued

In 2002, the Grassland National Park Management Plan identified grazing as an important ecosystem function for maintaining mixed grass prairie and necessary to restore ecological integrity. The management plan proposed implementing a grazing prescription to represent regimes more consistent with historical patterns of migratory and sedentary bison herds. In 2005, the park prepared environmental impact assessment screening reports for two large-scale experiments: a grasslands grazing experiment and the reintroduction of plains bison. The grasslands grazing experiment was to involve the introduction of livestock to the park as part of a broader adaptive management process for restoring ecological integrity. The primary objective of the grazing experiment was to determine how grazing intensity alters spatial and temporal heterogeneity of the mixed-grass prairie community. Over a 10-year period, approximately 2,664 hectares of native prairie was to be set aside for the experiment, which was based on a 'before-after-control-impact' design process. The bison reintroduction project would involve an initial introduction of 50 to 75 animals to the park, transported from Elk Island National Park in Alberta. The long-term stocking rate objective is to develop a bison herd size of up to 358 animals over approximately 185 square kilometres.

Both EIAs were screening assessments prepared under the requirements of the Canadian Environmental Assessment Act. As the proponent of both projects, Grasslands National Park was also responsible for the design and implementation of a follow-up program. The two projects were different from most development initiatives in that their overall purpose was to introduce disturbance to the environmental system in order to learn how it responds to particular management actions. At the same time, however, a number of unintended, potentially adverse environmental effects were also recognized.

A total of 45 potentially adverse impacts were identified in the impact statements, as well as 64 prescribed mitigation measures. Impacts and mitigation concerned a range of biophysical and socio-economic issues, including visitor experience, vegetation, wildlife, aquatics, and cultural resources. An integrated follow-up program, encompassing both projects, was designed in 2005 and focused on following-up for mitigation effectiveness. The objectives of the follow-up program were to: i) ensure that mitigation measures identified in the impact statements were implemented; ii) establish systems and procedures for this purpose; iii) monitor the effectiveness of mitigation; and iv) take any necessary action when unforeseen impacts occurred or when mitigation measures were not performing as expected.

The follow-up framework consisted of four core components: i) an implementation audit or one-time check or verification that proposed mitigation measures had been implemented; ii) compliance monitoring and regular inspection to ensure that agreed-upon impact management procedures and best-management practices identified in the impact statements were being adhered to; iii) effects monitoring and evaluation to track and evaluate changes in a range of biophysical, economic, and social variables so as to compare data collected during project implementation and operation against baseline conditions; and iv) the development of effectiveness indicators, targets, and thresholds so as to determine whether the mitigation measure was working to avoid or minimize potentially adverse effects. The first year of the follow-up program was implemented in 2007.

Source: Noble 2006.

Follow-up Components

Environmental impact assessment follow-up can be thought of as consisting of three interrelated components: monitoring, auditing, and ex-post evaluation. Generally speaking, **monitoring** is an activity designed to identify the nature and cause of change. More specifically, it is a data collection activity undertaken to provide specific information on the characteristics and functioning of environmental and social variables (Bisset and Tomlinson 1988). Monitoring usually consists of a program of repetitive observation, measurement, and recording over a period of time for a defined purpose (Arts and Nooteboom 1999), the objective of which is to detect whether change in a particular variable has occurred and to estimate its magnitude. Determining the causes of change and whether such changes are the consequence of the project is an essential part of monitoring.

Originating in economics and accountancy, **auditing** refers to objective examination or a comparison of observations with predetermined criteria. There are a variety of types of audits (Box 10.2), but auditing generally is a periodic activity that involves comparing monitoring observations against a set of criteria, such as standards or expectations, and reporting the results. Whereas monitoring is often a frequent or continual process, auditing is usually a periodic or single event. There is little point in collecting monitoring data unless those data are subject to some form of comparative analysis or audit (Arts and Nooteboom 1999).

A third and related concept is **ex-post evaluation**, which refers to the collection, structuring, analysis, and appraisal of information concerning project impacts and making decisions on remedial actions and communication of the results of this process (Arts 1998, 75). The differences and relationships among these three components are illustrated in Figure 10.2.

RATIONALE FOR POST-DECISION MONITORING

There are several, often overlapping reasons for undertaking monitoring activities. However, three broad reasons for post-decision monitoring can be identified: monitoring for compliance, monitoring progress, and monitoring for understanding.

Box 10.2 Types of EIA Audits

Draft EIS audit: Review of the project EIS according to its terms of reference.

Project impact audit: Determination of whether the actual environmental impacts of the project were those that were predicted.

Decision-point audit: Examination of role and effectiveness of the EIS based on whether the project is allowed to proceed and under what conditions.

Implementation audit: Determination of whether the recommendations included in the EIS were actually implemented.

Performance audit: Examination of the proponent's environmental management performance and ability to respond to environmental incidents.

Predictive technique audit: Comparison between actual and predicted effects of the project.

Source: Tomlinson and Atkinson 1987.

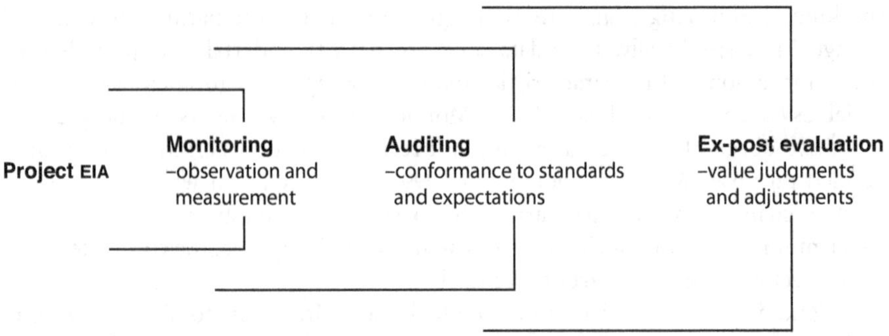

Figure 10.2 Follow-up components.

Source: Based on Arts 1998.

Compliance Monitoring

The primary purpose of **compliance monitoring** is to determine project compliance with regulations, mitigation commitments, agreements, or legislation. In this sense, follow-up has a *control* function—to ensure that a project is operating within specified guidelines. Compliance monitoring alone does not fulfill the requirements of an EIA follow-up program. It is simply a means of ensuring that what a proponent said would be done in the EIS has actually been done once the project has been implemented. There are several types of compliance monitoring, including inspection monitoring, regulatory permit monitoring, and agreement monitoring.

Inspection monitoring, the simplest form of monitoring, is site-specific and involves checking to ensure that operating procedures are being followed and that environmental degradation is not occurring. Inspection monitoring typically involves on-site visits and regular reporting of relevant activities and is mostly used for regular checking for compliance with agreed-upon procedures and operation within acceptable standards of safety.

Regulatory permit monitoring is also site-specific and involves the regular documentation of conditions required for maintenance or renewal of a permit, such as permits for the operation of waste disposal systems.

The **monitoring of agreements** between project proponents and affected groups, such as impact benefit agreements, are becoming commonplace, particularly in Canada. They sometimes include a monitoring component to track changes in population, housing, and other infrastructure demands in order to assign costs associated with the project and to ensure compliance with stated commitments.

Progress Monitoring

The purpose of progress monitoring is to confirm anticipated outcomes and to alert managers to unanticipated outcomes. In this sense, follow-up has a *watchdog* function and allows managers to measure project and environmental progress and to respond to adverse environmental change in a timely fashion when necessary. Common approaches to progress monitoring include ambient environmental quality monitoring, monitoring for management, cumulative effects monitoring, and project evaluation monitoring.

Ambient environmental quality monitoring is concerned with the effects of the project on its surrounding environment. Information collected prior to project approval and implementation at and near the site and at control sites can provide baseline data against which to compare data collected during the development, operation, and post-project phases. While most ambient environmental quality monitoring is associated with measures of the status of biophysical phenomena such as air and water quality, socio-economic changes such as quality of life or health and well-being can also be the focus.

Monitoring for management may include tracking and evaluating changes in a range of environmental, economic, and social variables. This type of monitoring is usually associated with high-profile projects with uncertain outcomes, which have the potential to result in significant adverse outcomes unless prompt action is taken to address issues as they emerge. The size and scope of the monitoring requirements may be such that the co-ordination of the monitoring program is undertaken by a formally constituted and funded monitoring organization, such as the Independent Environmental Monitoring Agency (IEMA) established to oversee outcomes from the Ekati Diamond Project in northern Canada (www.monitoringagency.net) and to ensure that the project meets the requirements set forth in IEMA's environmental approval. Monitoring for management is the primary focus of its mandate, and finding solutions to environmental management issues arising from the project is a primary objective.

Cumulative effects monitoring is less site- and project-specific and attempts to monitor the accumulated effects of developments within a particular region. The broad range of interests involved and the need for co-ordination mean that cumulative effects monitoring is best achieved by an organization mandated with monitoring responsibilities. The nature of cumulative environmental effects is discussed in greater detail in Chapter 12.

Also referred to as productivity measurement, **project evaluation monitoring** is concerned with measuring a project's performance with respect to established goals or objectives, such as overall efficiency. The focus is often social or economic in nature and includes measures of performance of programs designed to provide job training to address social concerns.

Monitoring for Understanding

A third reason for monitoring is to better understand the complex relationships between human actions and environmental and social systems. In this sense, follow-up has a *learning* function by increasing knowledge and understanding that can be applied to the science of assessment of future projects or related policy decisions. Monitoring for understanding includes experimental monitoring and monitoring for knowledge.

The purpose of **experimental monitoring** is to generate information and knowledge about environmental systems and their impacts through research methodologies that test specific hypotheses. Whereas the approaches discussed above, while designed to monitor changes, do not take anticipated consequences into account, experimental monitoring is guided by questions to test specific hypotheses. In this sense, experimental monitoring is science-driven rather than motivated by impact management per se.

Monitoring for knowledge is a type of data collection and reporting that often takes place well after impacts occur. Rather than being used for impact management purposes, the data are used to provide insights for the management of future projects (Box 10.3).

REQUIREMENTS FOR EFFECTIVE FOLLOW-UP

In Canada and elsewhere, post-decision follow-up has been described as less than satisfactory. While weak or non-existent legal requirements and institutional support mechanisms may in part be contributing to the current state of practice, a number

Box 10.3 Monitoring for Knowledge: The Alma Smelter, Quebec

The Alma smelter began operation in 2001 in Alma, Quebec. It employs 865 people and has a capacity of 407,000 metric tons of aluminum ingots. Researchers at the University of Quebec at Chicoutimi undertook a five-year study between 1998 and 2003 to document the social impacts of the project on Alma, which has a population of 30,126. The study was carried out in three parts in 1998, 2000, and 2002, corresponding to the planning, construction, and operating phases of the project. The study examined the quality of life of local residents, using a series of subjective indicators to reflect perceptions and degree of satisfaction with living conditions.

The stated medium- and long-term research objectives of the research were to:

- develop an impact follow-up model of Alma's aluminum smelter before 2001;
- generate applied knowledge concerning:
  - economic characteristics of study zones,
  - citizen perceptions concerning quality of life and the environment,
  - expected social and economic impacts before, during, and after construction of the smelter,
  - assessment methods and impact follow-up models,
  - processes, modes, and analysis of public and community participation;
- set up a geo-referenced database for urban and environmental planning and management;
- propose, if needed, preventive and mitigation measures during the plant's set-up;
- generalize results with regard to the establishment of future electrolysis plants in Quebec;
- allow development of appropriate expertise in the field of environmental monitoring and regional development.

Quality of life, based on a related concept of health proposed by the World Health Organization, was defined as 'a state of complete physical, psychological, and social well-being'. The approach used was to measure a series of subjective indicators, reflecting residents' perceptions of and degree of satisfaction with living conditions, through face-to-face interviews with members of the community. The same questionnaire was used in each of the three phases of the study to survey a stratified sample of the population by census enumeration area. Response rates ranged between 61 per cent and 73 per cent.

The study indicated that as stakeholders, residents were generally satisfied, despite an increase in the impacts during the construction phase of the smelter. Responses

continued

gathered showed a broad consensus on a number of issues. An Alma resident is generally someone who:

- is satisfied with his/her quality of life with respect to the biophysical and community environment as well as the economy;
- is concerned about issues related to health and family;
- sees Alma as a dynamic community and does not plan to move;
- perceives few environmental risks, while remaining aware of the possible effects of industrial activity;
- is somewhat dissatisfied with the traffic situation and noise caused by blasting during the construction phase.

Follow-up for impact management purposes did not appear to be a primary focus of the research, and there is no information to suggest that preventive or mitigative measures were identified, proposed, or acted upon with respect to the project. Rather, the emphasis was on using follow-up as a database and learning tool, providing information and experience that could be applied to other, similar projects.

Source: From Storey and Noble 2004. For additional details, see the research project website at www.uquac.ca/~msiaa.

of substantive and procedural elements are also required to facilitate effective post-decision EIA follow-up programs.

Objectives and Priorities Identified

Perhaps the most important requirement for effective follow-up is the clear articulation and identification of follow-up and monitoring program objectives and priorities at the outset of program design. If a clear set of objectives for follow-up and monitoring do not exist, whether for monitoring for compliance, performance, or knowledge, then one cannot adequately determine whether the follow-up or monitoring program itself is effective. The specific objectives of follow-up and monitoring programs vary from one project to another (Box 10.4) and may include such objectives as verifying impact predictions or ensuring regulatory compliance for project licensing. Failure to state clear program objectives and priorities leads to dissatisfaction with and confusion regarding what the monitoring program is actually trying to achieve. Consequently, consideration should be given to follow-up and monitoring objectives and priorities from the outset of EIA—during the scoping process, when key environmental issues are first identified, since that will allow pre-development baseline monitoring. Specific monitoring program design can be postponed until the post-decision stage when project design elements are finalized and priorities for management purposes can be better determined.

Targeted Approach to Data Collection

Not everything of interest or importance in an EIA can be monitored and effectively followed up. Further, because of the complexity of most stressor-response relationships, it is impossible to characterize all of the variables. Therefore, monitoring data, particularly for the biophysical environment, must be targeted and focused on the

Box 10.4 Follow-up Program Objectives for Two Canadian Mining Projects

Voisey's Bay nickel mine-mill project

1. To continue to provide baseline data so that project activities can be scheduled or planned to avoid or reduce conflict with VECs.
2. To verify earlier predictions and evaluate the effectiveness of mitigation to lower uncertainty or risk.
3. To identify unforeseen environmental effects.
4. To provide an early warning of undesirable change in the environment.
5. To improve understanding of environmental cause-and-effect relationships.

Ekati diamond mine project

1. To ensure regulatory compliance.
2. To measure operational performance and the effectiveness of mitigation strategies.
3. To monitor natural environmental changes as well as those caused by the project (environmental effects monitoring).
4. To assess the validity of impact predictions.
5. To trigger response to and mitigation of unexpected adverse effects.

indicators that are useful for understanding, or at least correlating, stressor–response relationships. Such indicators must be comparable over space and time, and in order to differentiate project-induced change from natural change, these variables and indicators and the data derived from them must be comparable to previous, current, and forecasted baseline conditions.

One approach to indicator monitoring is to focus on early warning indicators. Measurable parameters of either a non-biological or a biological nature, **early warning indicators** serve to indicate stress on particular VECs before these VECs themselves are adversely affected. Early warning indicators might include, for example, nitrogen and phosphorous concentrations as indicators of water quality or worker productivity and number of sick days taken as early warning indicators of worker stress (Table 10.1). Early warning indicators must be:

- directly or indirectly related to the VEC;
- physically possible to monitor;
- amenable to quantitative analysis.

Hypothesis-Based or Threshold-Based Approaches

For each variable to be monitored, significance levels and probability levels must be specified as part of the monitoring program protocol. Successful impact monitoring requires that such information be formulated as testable hypotheses; analyses of significance will have meaning only when they are made against an a priori null hypothesis (Bisset and Tomlinson 1988). For each indicator, appropriate thresholds should thus be established as a measure of environmental effects. Such thresholds may be established on the basis of:

- consultation with regulatory agencies;
- levels of acceptable change;
- scientific research or recommendation;
- project goals or objectives.

This approach, which is common to many biophysical environmental effects monitoring (EEM) programs, is often based on testing to determine whether changes in specific indicators exceed stated threshold levels. Monitoring of these indicators helps to verify that the management measures implemented are effective, but since there is no attempt to determine the specific level of the effect, predictive accuracy is not the primary focus of the assessment (Storey and Noble 2004). Such an approach was adopted for the Hibernia offshore oil platform construction project in Newfoundland (Box 10.5).

Table 10.1 Selected Environmental Components, VECs, and Monitoring Indicators for the Cold Lake Oil Sands Project

Environmental component	*Issues of concern*	*Valued components*	*Warning indicators*
Air systems	Acidic deposition, odours, greenhouse gas emissions	Air quality	Emitted gases transported over long distances (NOx, SO_2)
Surface water	Lowering of lake water levels, contamination of water	Water quality and quantity	Combined water volume withdrawals, water quality constituents affecting drinking water standards
Groundwater	Depletion of aquifers	Potable well water	Combined water volume withdrawals
Aquatic resources	Contamination of fish, increased harvest pressures	Sport fish species	Northern pike populations/health
Vegetation	Loss of vegetation through land clearing, effects of airborne deposition	Vegetation ecosites	Low bush cranberry, aspen, white spruce
Wildlife	Loss, sensory alienation and fragmentation of habitat, direct mortality due to increased traffic and hunting harvest	Hunted and trapped species	Moose, black bear, lynx, fisher populations/health

Source: Based on Hegmann et al. 1999.

Box 10.5 Hibernia Offshore Platform Construction Project, Biophysical Environmental Effects Monitoring (BEEM) Program

The Hibernia offshore oil field was discovered on the Grand Banks of Newfoundland in 1979. Its proposed development was subject to a panel review under the Canadian Federal Environmental Assessment Review (FEARO) process. Approval for the project was granted in 1986, and development began in 1990. The Hibernia Management and Development Company (HMDC), in conjunction with relevant agencies forming part of the Hibernia Construction Sites Environmental Management Committee, established a multi-year (1991–6) program to monitor the effects of the Hibernia Gravity Base Structure (GBS) construction site on the marine environment of Bull Arm, Trinity Bay. Survey data on benthic fauna, marine fish and shellfish habitat, chemical and biological characteristics of the water column, sediments and resident benthic fauna, and fisheries utilization had been collected as part of the assessment process, which provided both a baseline for subsequent monitoring activity and information for determining various monitoring criteria.

The construction site at Great Mosquito Cove, Newfoundland, and the adjacent deepwater site were considered the major potential point sources of effluent and/or accidental discharge. The area is rich in marine invertebrates, fish, and marine mammals, and Bull Arm had traditionally supported a diverse inshore commercial fishery, which formed the economic base of many of the communities in the area. The general objectives of the biophysical environmental effects monitoring (BEEM) program included an assessment of the effectiveness of environmental protection and mitigation measures but also included providing early warning of undesirable change and assurance that impacts predicted to be insignificant were in fact insignificant ('comfort monitoring').

Selection of monitoring variables to reflect significance to the environment and management needs was based on the following criteria: (1) the variable is of social, economic, or biological importance; (2) potential exists that the variable would be affected by the project; (3) species are common/accessible in the area; (4) species are amenable to scientific sampling and monitoring; (5) baseline data are either not required or easily collected; (6) variables can be monitored over time for relatively low cost.

Variables chosen to reflect the specific objectives of the program included: (1) mixed function oxygenase (MFO) activity and bile metabolite level in winter flounder; (2) gill deformity in winter flounder; (3) trace metal and hydrocarbon levels in blue mussel tissue; (4) trace metal and hydrocarbon levels in shallow and deepwater sediments; and (5) nearshore sedimentation rates (LGL 1993).

For each measured variable, an impact (null) hypothesis was developed, stating that 'activities associated with site development and construction of the GBS and Topsides in Bull Arm will not elevate the concentration or degree of the variable to a level which exceeds the maximum allowable effects level (MAEL) for that variable.' The MAELs for the variables were based on values from different sources. For example, those applying to mussels and sediments were drawn from the Canadian Fish Health Inspection/Food and Drug Act Regulations, the Food and Agriculture Organization of the United Nations, CEPA Ocean Dumping Regulations, and Canadian Marine Environmental Quality Guidelines. For variables without values from these sources, other comparable local data and professional opinion were used to define MAELS (LGL 1993).

Field samples of mussels, flounder, and sediments were collected and prepared for analysis by consultants to HMDC. Established commercial and university laboratories undertook

continued

analysis of the samples, and the results were summarized and reported annually (LGL 1993–7). The findings from the analyses were that none of the null hypotheses developed for the BEEM program could be rejected. In other words, the construction project did not have impacts on the marine environment beyond acceptable levels. A similar approach was adopted for the environmental effects monitoring program for the offshore development phase of the Hibernia project, and subsequent offshore developments in the area—Terra Nova in 2002 and White Rose in 2005—adopted similar monitoring programs.

Source: Storey and Noble 2004.

Effects-Based Approaches

The focus of EIA is often on predicting the potential effects of a proposed development on specified VECs using a stressor-based approach. In other words, the nature and magnitude of changes in a particular VEC or VEC indicator are predicted on the basis of the conditions imposed on the local environment by project development (Kilgour et al. 2007). Under such an approach, it is assumed that there is an adequate level of understanding of the properties of the affected components to identify and predict the most likely outcomes. For simple projects and environmental systems, the use of a stressor-based approach and subsequently monitoring the stressors may be an appropriate means of effectively assessing the actual effects of development on the receiving environment. The problem, however, is that projects and environmental systems are rarely that simple, and focusing monitoring programs solely on stressor–VEC response relationships may mask broader systemic environment changes caused by the project if no measurable changes are detected in the VECs themselves.

Essential to follow-up and monitoring programs then is a complementary effects-based approach that focuses on the performance of the environmental system and VEC indicators rather than solely on the stressors and individual VEC responses. In Canada, Kilgour et al. (2007) explain, such effects-based monitoring programs are used by many agencies to determine whether environmental quality has been compromised, and they are currently a legal component under the Fisheries Act for monitoring and evaluating the effects associated with discharges to aquatic systems from metal mines and pulp mills. Among the current challenges in developing and implementing such effects-based monitoring systems in EIA practice, however, is that EIA and environmental effects monitoring practitioners have generally operated independently and 'have evolved relatively distinct jargon that is not easily transferred between the disciplines' (Kilgour et al. 2007). Adding to this complexity, not all VECs identified and assessed in an EIA are easily tied to specific measurable parameters or indicators in the broader environmental system that can be effectively monitored. If an effect cannot be tied to either a stressor or a VEC identified in a project assessment, challenges emerge concerning roles and responsibilities for mitigation.

Control Sites

Where possible, **control sites** should be established as reference monitoring locations to compare with the treatment (project-affected) locations. This will help in correctly differentiating between project impacts and natural change and will facilitate the spatial and temporal consistency of the effects monitoring program with initial project impact predictions. In cases where a well-established control site is not readily available, it may be possible to create an artificial one using a **gradient-to-background monitoring** approach.

This approach assumes that there is a well-defined, localized source of impact, such as pollution, and that effects can be monitored at increasing distances from the point source. With increasing distance from the point source, effects should decrease and eventually reach background or ambient conditions. The point on this gradient at which background levels are attained is considered the control site (Figure 10.3).

Continuity

There should be continuity in data collection and handling procedures to ensure that monitoring data are transferable and comparable. Failure to achieve continuity may lead to problems in comparing monitoring results and the significance of those results from one monitoring period to the next. Such quality control problems in the continuity of data collection were evident in the Rabbit Lake uranium mine project in Saskatchewan. After decades of biophysical monitoring and data collection, the proponent was unable to make a direct connection between project actions and the impacts of project-induced environmental change (Box 10.6).

Adaptability, Flexibility, and Timeliness

Care should be taken to ensure that as project or environmental conditions change, data quality is consistent and comparable over time. While objectives and indicators need to be established at the outset, project changes, unanticipated effects, emerging project concerns, and indicator choices may warrant changes in elements of the monitoring as it proceeds (Davies and Sadler 1990). This requires that monitoring and reporting be completed in a timely fashion so that those using the results can respond promptly. While data collected late in the process are still valuable for reflection and learning, such monitoring processes are often a waste of time from the perspective of impact management.

Inclusiveness

Follow-up programs should address not only traditional biophysical impacts but socio-economic and other human and cultural impacts as well. In practice, however, social issues have been given less than sufficient attention in post-decision monitoring. While the assessment of socio-economic and other effects on the human environment

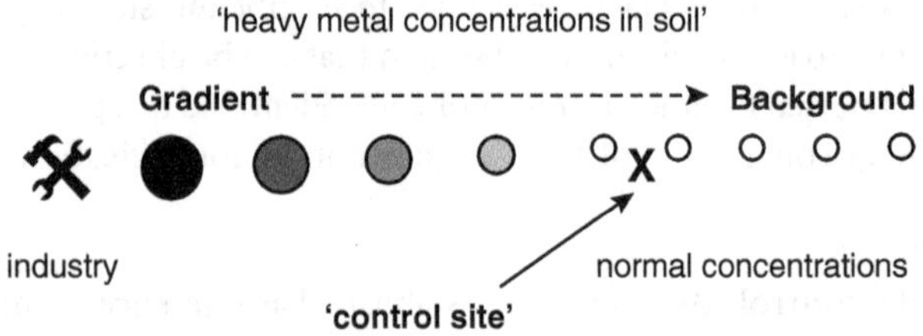

Figure 10.3 Gradient-to-background approach.

Box 10.6 Continuity in Monitoring Data: The Rabbit Lake Uranium Mining Project

The Rabbit Lake mining project located in northern Saskatchewan is the province's oldest operating uranium mining and milling facility. Since the discovery of deposits in the area in 1968, the Athabasca Basin of northern Saskatchewan has been the world's premier exploration region for high-grade uranium deposits. In 1987, subsequent exploration activity on the Rabbit Lake mine site identified several additional radioactive occurrences, and the proponent submitted an EIS to federal and provincial regulatory agencies for approval to mine three new ore bodies. The project was approved and a licence issued for development.

Four years later, a joint federal-provincial EIA panel was appointed to examine the environmental, health, and socio-economic effects of uranium mining activities in northern Saskatchewan, including the Rabbit Lake mine and extension. The panel's report, released in 1993, identified contamination of the biophysical environment and subsequent exposure pathways to radionuclides and heavy metals of primary concern. In its presentation to the panel, the proponent noted that it had collected baseline data and monitored the biophysical environment near the mine site since the late 1970s. However, the panel noted that while monitoring requirements that met regulatory requirements were in place and data collection was ongoing, there was some concern over the quality of the monitoring data, consistency of methods used to test for radionuclides and trace elements, and effectiveness of the monitoring program in determining the impacts of mining activities.

Monitoring and testing procedures had changed several times throughout the 1980s, and data collected during 1989 and 1990 were discarded because of quality problems. After more than a decade of biophysical monitoring and data collection, there were few comparable data concerning the effects of radionuclides from mining operations on fish— a valuable resource base for northern residents.

are often given as much attention as biophysical effects in impact statements and panel reports, such attention is often not carried over into follow-up—and this is particularly true for the 'social' components of socio-economic impacts (Box 10.7). When social and other human issues do carry over to the follow-up stage, they tend to be treated with considerably less rigour than biophysical components (Box 10.8). Proponents often wish to distance themselves from socio-economic follow-up, partly because of the complexity of the pathways that link project actions to socio-economic impacts. In the case of the Cluff Lake uranium mine in northern Saskatchewan, the final panel report of the Cluff Lake Board of Inquiry (1978, 174) explicitly recognized the difficulty of assessing the social and other human impacts associated with uranium mining activities, stating:

> there now exists in the north (and it has nothing to do with uranium mining) a social disorder. . . . To superimpose upon that kind of society a project such as a uranium mine and mill which has the potential of exacting additional social costs and then try and measure those additional costs presents a near impossible task.

In principle, according to discussions with an industry representative, there are often so many confounding factors in the communities that it is impossible to tell whether

Box 10.7 Canadian Experiences with Socio-economic Monitoring: Confederation Bridge Project

In May 1989, Public Works Canada submitted a Bridge Concept Assessment document for review. The proposed project was the construction of a 13-kilometre bridge over the Northumberland Strait, extending from Cape Tormentine, New Brunswick, to Borden, Prince Edward Island. The mandate of the EA panel reviewing the proposal was to examine the environmental and socio-economic effects, both beneficial and harmful, of the proposed bridge. There was consensus on the need for improved transport service between Prince Edward Island and New Brunswick, but the panel's recommendation was that the risk of harmful effects, particularly concerning delays in spring ice retreat, risk of damage to the ecosystem, and the displacement of more than 600 Marine Atlantic ferry workers, was such that the project should not proceed.

The panel's recommendation was not accepted by the federal government, and following a 1990 study by an independent Ice Committee, the minister of environment approved the project based on the committee's conclusion that the designs for the bridge under consideration would meet the panel's ice-out criteria. Construction began in the fall of 1993, and the Confederation Bridge was completed in the spring of 1997. An Environmental Management Plan (1993) was prepared, which included a long-term Environmental Effects Monitoring (EEM) Program. In line with the CEAA definition of follow-up and the emphasis on the physical environment, the goals of the EEM were to evaluate the effectiveness of environmental protection procedures and to verify predictions regarding the potential biophysical effects of the project. Ice characteristics were initially documented in 1993 before the bridge was constructed, studied throughout project construction, and monitored after the bridge was completed. Notwithstanding the concern expressed in the initial panel report over the issue of displacement of ferry workers, there was no formal attempt to follow up on this or any other of the potential social or economic effects raised in the assessment document, at the public hearings, and by the panel.

Source: Storey and Noble 2004.

Box 10.8 Canadian Experiences with Socio-economic Monitoring: Ekati Diamond Mine

In 1994, the Canadian Department of Indian Affairs and Northern Development initiated an environmental review of Canada's first diamond mine, 300 kilometres northeast of Yellowknife in the Northwest Territories. The proposal involved the development of a diamond mine in an area of unsettled and overlapping Aboriginal land claims where there had been little previous industrial development. The proponent, now BHP Billiton (BHPB), submitted its assessment documents in 1994, and a full panel review followed.

The importance of follow-up was recognized, and a variety of suggestions were made to ensure that it was undertaken. The panel, based on information from the government of the Northwest Territories (GNWT) that it had already identified and collected data on a number of health and wellness indicators, recommended a partnership approach, primarily between the proponent and the GNWT.

continued

In 1996, a Socio-economic Agreement was signed between the GNWT and BHPB for the Ekati project. The agreement is intended to promote the development and well-being of the people of the Northwest Territories, particularly people in communities close to the mine. The agreement focuses on monitoring and promoting social, cultural, and economic well-being. The GNWT is responsible for the establishment and maintenance of the monitoring program. The objective is to look for ways to strengthen opportunities and to mitigate negative effects associated with the project. Public statistics and surveys of mine employees are the primary data sources used to track a series of indicators chosen to reasonably match the possible effects identified during the assessment phase for the project. In addition to data on social stability and community wellness, BHPB issues its own reports describing realization of business and employment opportunities and also reports directly to the Aboriginal groups with whom it has signed impact benefits agreements.

Public statistics for 14 indicators are currently used to monitor and assess the effects of the Ekati project. Data are monitored and reported for the small communities in the West Kitikmeot Slave area, Lutselk'e, Rae-Edzo, Rae Lakes, Wha Ti, Wekweti, Dettah, and Ndilo, as well as for Yellowknife. These data are then compared with information for the rest of the Northwest Territories. The results are presented in an annual report. While these data provide indications of overall changes within the region for the indicators tracked, most of the results are at best inconclusive in linking social and economic changes in the communities to the diamond project activities. Cause and effect are always difficult to determine for social variables, but the 'coarseness' of the indicators used to measure change, the small size of some communities, and the quality of some of the available data make it particularly difficult to do so in this case.

Key socio-economic monitoring indicators for the BHPB Ekati project include:

Social stability and community wellness

- number of injuries
- number of potential years of life lost
- number of suicides
- number of teen births
- number of children in care
- number of complaints of family violence
- number of alcohol/drug-related crimes
- number of property crimes
- number of communicable diseases
- housing indicators

Non-traditional economy

- average income of residents
- employment levels and participation
- number of income assistance cases
- high school completion
- cultural well-being

Sources: GNWT 2001; Storey and Noble 2004.

or not there has been an effect. Further, many communities are too far away to permit monitoring direct effects based on scientific risk and pathways monitoring.

If sustainable development is the objective of EIA, the biophysical and socio-economic components must be given equal consideration throughout all phases of EIA, including post-decision monitoring. That said, one question that emerges from recent practice experience is: Who is responsible for **socio-economic monitoring**? Most proponents suggest that they are willing to cooperate with government and other bodies by sharing project information but maintain that socio-economic follow-up is primarily the responsibility of other parties. Mobil Oil held to this view in connection with the Hibernia offshore oil project. Similarly, in the more recent Voisey's Bay mine-mill project, the proponent took the position that the financial provisions in the impact benefits agreements to be signed with the Labrador Inuit Association and the Innu Nation were in part intended to provide these groups with the resources to carry out any socio-economic follow-up studies that they deemed necessary (Voisey's Bay Mine and Mill Environmental Assessment Panel 1999).

MONITORING METHODS AND TECHNIQUES

As with impact prediction, there is no single set of monitoring techniques for all projects and components. The type of monitoring technique selected to provide data and to evaluate environmental change depends on the nature of the environmental components, the purpose of the data, and particular monitoring program objectives. Thus, for both biophysical and socio-economic monitoring, a variety of quantitative and qualitative techniques, or combinations thereof, are available. For example, socio-economic impacts can be monitored using annual surveys of residents' perceptions of quality of life and local police and hospital records. Selected examples of techniques for monitoring biophysical components are listed in Table 10.2. When selecting a technique for monitoring, as for any other EIA procedure, it is important to keep in mind the nature and resolution of the data required.

Table 10.2 Biophysical Monitoring Components, Parameters, and Techniques

Components	*Parameters*	*Techniques*
wildlife habitat	fragmentation	remote sensing aerial photography
rangeland health	vegetation cover vegetation composition soil structure	remote sensing field transects trend transects
water quality	metals nutrients physical parameters	stream gauges and go-flow samplers chemical analysis
lake biology	lake benthos	Eckman Grab stratified depth sampling and species counts
stream biology	stream benthos fish	Hester-Dendy sampler and frequency counts tagging electro fishing age-length keys
hydrology	water level	staff gauges pressure transducers

KEY TERMS

ambient environmental quality monitoring
auditing
compliance monitoring
control site
cumulative effects monitoring
decision-point audit
draft EIS audit
early warning indicators
experimental monitoring
ex-post evaluation
gradient-to-background monitoring
implementation audit
inspection monitoring
life-cycle assessment
monitoring
monitoring for knowledge
monitoring for management
monitoring of agreements
performance audit
predictive technique audit
project evaluation monitoring
project impact audit
regulatory permit monitoring
socio-economic monitoring

STUDY QUESTIONS AND EXERCISES

1. What provisions exist for post-decision monitoring and auditing under your national, provincial, or state EIA system? Are these provisions mandatory or voluntary?
2. What is the value added to EIA from follow-up and monitoring activities?
3. Who should be responsible for monitoring the environment after project approval?
4. Why are socio-economic effects difficult to monitor post–project implementation? How might we address these difficulties?
5. What is the role of the public in environmental monitoring?
6. Obtain a completed project EIS from your local library or government registry, or access one on-line. Is there a monitoring component to the impact statement? What environmental aspects are included in the monitoring program? Are methods and techniques for monitoring identified? Compare your findings with those of others.

REFERENCES

Arts, J. 1998. *EIA Follow-up: On the Role of Ex-post Evaluation in Environmental Impact Assessment.* Groningen, Netherlands: Geo Press.

Arts, J., P. Caldwell, and A. Morrison-Saunders. 2001. 'Environmental impact assessment follow-up: Good practice and future directions'. *Impact Assessment and Project Appraisal* 19 (3): 175–85.

Arts, J., and S. Nooteboom. 1999. 'Environmental impact assessment monitoring and auditing'. In J. Petts, Ed., *Handbook of Environmental Impact Assessment.* London: Blackwell Science.

Bisset, R., and P. Tomlinson. 1988. 'Monitoring and auditing of impacts'. In P. Wathern, Ed., *Environmental Impact Assessment: Theory and Practice.* London: Unwin Hyman.

Canadian Environmental Assessment Agency. 2007. *Follow-up Programs under the Canadian Environmental Assessment Act.* Operational Policy Statement. Ottawa: Canadian Environmental Assessment Agency.

Cluff Lake Board of Inquiry. 1978. *Cluff Lake Board of Inquiry Final Report.* Regina.

Davies, M., and B. Sadler. 1990. *Post-project Analysis and the Improvement of Guidelines for Environmental Monitoring and Audit*. Report EPS 6/FA/1 prepared for the Environmental Assessment Division, Environment Canada. Ottawa.

GNWT (Government of the Northwest Territories). 2001. *Communities and Diamonds: Socio-economic Impacts on the Communities of Lutselk'e, Rae-Edzo, Rae Lakes, Wha Ti, Wekweti, Dettah, Ndilo and Yellowknife*. Yellowknife: Annual Report of the Government of the Northwest Territories under the BHP Socio-economic Agreement.

Kilgour, B., et al. 2007. 'Aquatic environmental effects monitoring guidance for environmental assessment practitioners'. *Environmental Monitoring and Assessment* 130 (1–3): 423–36.

LGL. 1993–7. *The Hibernia GBS Platform Construction Site Marine Environmental Effects Monitoring Program. Years 1–5*. Report prepared for the Hibernia Management and Development Company Ltd by LGL Ltd, St John's.

Noble, B. 2006. *Environmental Impact Assessment Follow-up Prescription and Reporting Framework: Grasslands Grazing Experiment and Reintroduction of Plains Bison, Grasslands National Park of Canada*. Val Marie, SK: Grasslands National Park.

Ramos, T., S. Caeiro, and J. de Melo. 2004. 'Environmental indicator frameworks to design and assess environmental monitoring programs'. *Impact Assessment and Project Appraisal* 22 (1): 47–62.

Storey, K., and B. Noble. 2004. *Toward Increasing the Utility of Follow-up in Canadian EA: A Review of Concepts, Requirements and Experience*. Report prepared for the Canadian Environmental Assessment Agency. Gatineau, QC: CEAA.

Tomlinson, P., and S. Atkinson. 1987. 'Environmental audits: Proposed terminology'. *Environmental Monitoring and Assessment* 8 (3): 187–98.

Voisey's Bay Mine and Mill Environmental Assessment Panel. 1999. *Report on the Proposed Voisey's Bay Mine and Mill Project/Environmental Assessment Panel*. Hull, QC: Canadian Environmental Assessment Agency.

CHAPTER 11

Public Participation in EIA

INVOLVING THE PUBLICS

Engagement of the public in impact assessment is required in some form in most EIA systems around the world. Broadly defined, public participation refers to the involvement of individuals and groups that are positively or negatively affected by a proposed intervention subject to a decision-making process or are interested in it (André et al. 2006). Public participation initiatives began to develop and have an influence on environmental decision-making during the 1960s in the United States. The public was a critical driving force behind the initial development of EIA, and since NEPA in 1969, public involvement has increasingly been recognized as a key element in EIA practice. Interestingly, however, under the NEPA system, requirements for public involvement in the EIA process are limited to public scoping of the issues to be addressed in assessment, review of the draft EIS, and court challenge.

There are international provisions with respect to public participation in EIA, including the 1991 Espoo Convention on Environmental Impact Assessment in a Transboundary Context and the 1998 Aarhus Convention on Access to Information, Public Participation in Decision Making and Access to Justice in Environmental Matters. Many national EIA systems also have specific requirements for public involvement in project evaluation and decision-making, with some more stringent than others. In the UK, for example, article 6 of Directive 85/337/EEC provides for public participation in EIA through the opportunity for a public review of the project's EIS before the project is initiated; there is no requirement for public involvement either before or during project planning. In Canada, one of the four main objectives of the Canadian Environmental Assessment Act is to ensure public participation in the EIA process. Revisions to the Act in 2003 strengthened the requirements for early public involvement during the screening phase and provided for incorporation of public and traditional knowledge during the assessment process. In practice, however, sustained involvement of the public in any substantive way, other than a review of the final EIS, is largely at the discretion of the project proponent and the responsible authority.

Rationale for Involving the Public

It is usually recommended that public involvement commence as early as possible in the EIA process, during the project scoping phase. In practice, however, public involvement is frequently limited to a public review of the completed project EIS. Some proponents have argued that it is more efficient to exclude the public from EIA, given that the public often lacks the project-specific expertise necessary to contribute to development decision-making, and argue instead for 'educating the public' rather than involving them. As a result, far too often the common answer to the question 'why involve the public in EIA?' is simply 'because it is required' (Shepherd and Bowler 1997). However, public involvement can be beneficial at all stages of the EIA process, from initial project design to post-decision analysis and monitoring (Table 11.1). The lack of public involvement during an EIA process, or inappropriate and tokenistic involvement, can be detrimental to the success of an EIA and to project approval (Box 11.1).

By involving the public in the decision-making process, it is possible to:

- define the problem more effectively;
- access a wider range of information, including traditional knowledge;

Table 11.1 Objectives of Public Involvement throughout the EIA Process

EIA stage	*Public involvement objectives*
Initial project design	• Early identification of affected interests and values • Identification of values relevant to site selection for conflict minimization
Screening	• Public review of decisions concerning EIA requirements • Early notification to affected interests of potential development
Scoping	• Further identification of active and inactive publics • Learning about public interests and values • Elicitation of local knowledge for baseline survey • Identification of potentially significant impacts • Identification of other areas of public concern and suggestion for management • Identification of alternatives • Establishment of credibility and trust between proponent and the publics
Impact prediction and evaluation	• Elicitation of values and knowledge to assist impact prediction • Identification of criteria for project evaluation • Development of public's technical understanding of project impacts
Reporting and review	• Informing public of project details, baseline conditions, likely impacts, and proposed management measures • Obtaining public feedback on key concerns, outstanding issues, and suggestions for improved management • Identification of errors or omissions in the EIS • Providing public an opportunity to challenge EIS • Assumptions and predictions
Decision-making	• Resolution of potential conflicts. • Final integration of responses from EIS review
Follow-up	• Maintenance of trust and credibility • Identification of management effectiveness • Elicitation of local knowledge in data collection and monitoring change

Source: Based on Petts 1999.

- identify socially acceptable solutions;
- ensure more balanced decision-making;
- minimize conflict and costly delays;
- facilitate implementation;
- reduce the possibility of legal challenge;
- promote social learning.

Public involvement in EIA can be seen as meeting a number of ends. While public involvement may extend the time needed during the initial project planning and scoping phases, this initial investment is usually returned later in the process because it minimizes or avoids conflict and facilitates project approval and implementation (Noble 2004). From a corporate perspective, Eckel, Fisher, and Russell (1992) agree, suggesting that consultation with different groups—regulators, shareholders, governments, and communities—will help to clarify expectations about environmental performance and boost a firm's reputation with the public.

NATURE AND SCOPE OF PUBLIC INVOLVEMENT

Petts (1999, 147) defines public involvement in EIA as:

> a process of engagement, where people are enlisted into the decision process to contribute to it . . . provide for exchange of information, predictions, opinions, interests, and values . . . [and] those initiating the process are open to the potential need for change and are prepared to work with different interests to develop plans or amend or even drop existing proposals.

Box 11.1 The Power of Public Involvement and Nuclear Fuel Waste Disposal in Canada

Current debates about the disposal of nuclear fuel waste in Canada date back more than 30 years when in 1978, under the Canada/Ontario Nuclear Fuel Waste Management Program, the governments of Canada and Ontario directed Atomic Energy of Canada Limited (AECL) to develop the concept of deep geological disposal of Canada's nuclear fuel waste (NFW). A subsequent joint statement issued in 1981 established that a disposal site would not be selected until the concept itself was publicly reviewed and approved by both governments.

In 1988, the concept, along with related issues concerning NFW, was referred for public review under the Federal Environmental Assessment and Review Process Guidelines Order, and in 1989 an independent environmental assessment panel, which would later become known as the Seaborn Panel, was appointed to develop guidelines for the assessment and to conduct a public review of the assessment document upon completion.

Terms of reference defining the review were released in 1989, and responsibility for the review process was given to an independent review panel administered initially by FEARO and later by CEAA. The panel's mandate was to review the disposal 'concept' rather than a specific project and location, the implementing agency of the concept was not defined, a broad range

continued

of policy issues were to be considered, and the public review was to span five provinces. The terms of reference directed the review in four areas: evaluation of the acceptability of the nuclear fuel waste disposal concept, including the burden on future generations, and a future course of action; comparison of the Canadian disposal concept to the approaches for nuclear fuel waste management adopted by other countries; a focus explicitly on nuclear fuel waste and the disposal concept; prohibition against discussing energy policy, nuclear plant operation and construction, and military applications. The terms of reference were not publicly negotiated; in essence, the scope of the assessment was predetermined.

The panel embarked on a series of scoping meetings to gather public input on the concept and to set the scope of AECL's impact statement. Public meetings were held in 1990 in 14 different communities. Draft guidelines were released for public review, and the final guidelines were presented to AECL in 1992. The panel directed AECL to consider ethical, moral, and social perspectives as equally important as the scientific and technical information. In a review of the terms of reference and guidelines for the AECL concept review and impact statement, Murphy and Kuhn (2001) note considerable debate among AECL and government and non-government agencies and an array of publics as to what could and should be considered in an assessment of a NFW disposal facility. Some contended that the panel was hampered by a very specific mandate and terms of reference; others suggested that the terms of reference were clear and that the entire process required further streamlining to restrict hearing submissions to those that conformed to the official mandate of the review.

AECL submitted its EIS in 1994, and a series of public hearings followed throughout 1995. The hearings were co-ordinated by the review panel and addressed three major areas of concern: the management of NFW within a broad societal context; a technological review of the assessment concept itself; and local perspectives on public safety and acceptability of the concept. The panel conducted its review in 16 communities across Saskatchewan, Manitoba, Ontario, Quebec, and New Brunswick and received more than 500 written submissions.

The public's response was for a broader consideration of alternatives, broader in scope than what was considered in the assessment. The terms of reference for the AECL NFW assessment were focused on a technical review of the proposed concept to the near-exclusion of alternative definitions and perspectives of what constitutes NFW management and the social and ethical aspects that NFW management might involve (Murphy and Kuhn 2001). The review panel, however, went beyond the scope of the terms of reference to ensure consideration of broader social issues and concerns regarding NFW management in Canada. Interestingly, the panel's report to government found that AECL's concept was technically safe but not publicly acceptable, partly because only one plan option, AECL's proposed concept, was considered. The panel's report contained numerous recommendations, including the creation of an independent organization to manage and co-ordinate all activities dealing with nuclear fuel waste in the long term and that such an organization would be subject to regulatory control and regular public review. In 2001, An Act respecting the Long-Term Management of Nuclear Fuel Waste was introduced. It passed in 2002 as the Nuclear Fuel Waste Act and resulted in the establishment of the nuclear Waste Management Organization—a private industry organization led by the major owners and producers of NFW, including provincial Crown nuclear power utilities and AECL. The Waste Management Organization has the mandate to review and select a preferred option for long-term NFW management. The Waste Management Organization was not the independent organization recommended by the panel, and the new Act and process have been criticized by some for lacking transparency and accountability.

Source: Noble and Bronson 2007.

Identifying the Publics

There is no such thing as 'the public' in EIA; rather, there are many 'publics'—some of whom may emerge at different times during the EIA process depending on their particular concerns and the issues involved. Mitchell (2002) makes a distinction between **active publics** and **inactive publics**. The active publics are those who affect decisions, such as industry associations, environmental organizations, quasi-statutory bodies, and other organized interest groups. Inactive publics are those who do not typically become involved in environmental planning, decisions, or issues and may include the 'average' town citizen (Diduck 2004). When involving the publics in EIA, it is important not to overrepresent the active publics and to ensure adequate representation of the inactive publics. Particular attention should thus be given to those who reside in the area where the project will be implemented and who may be directly affected by the project.

One way to approach identification of the publics for involvement in EIA is to consider the 'influence' of the public group versus the group's 'stake in the outcome' (Figure 11.1). Different publics may need to be involved in different capacities and at different points in the EIA process. For example, publics with little stake in the outcome (i.e., they will not be directly affected) and with limited power and influence over the project decision and EIA process may be considered spectators and involved in the EIA process only indirectly, through public communications, news releases, and education about the project. Individuals with a high stake in the outcome, such as the affected local population, but with limited power and influence, should be intimately involved throughout the EIA process so as to ensure that their concerns are addressed at the time of project decision; they have no influence over that decision. This second group is often referred to as the 'victims' of project development in that although they may experience some benefits from development, they also have the most to lose if project impacts are not properly managed. It is for this group that funding to participate, such as through the Canadian Environmental Assessment Agency's participant or intervener funding program, is important in order to facilitate participation and to ensure that the public has the necessary resources and capacity to become involved.

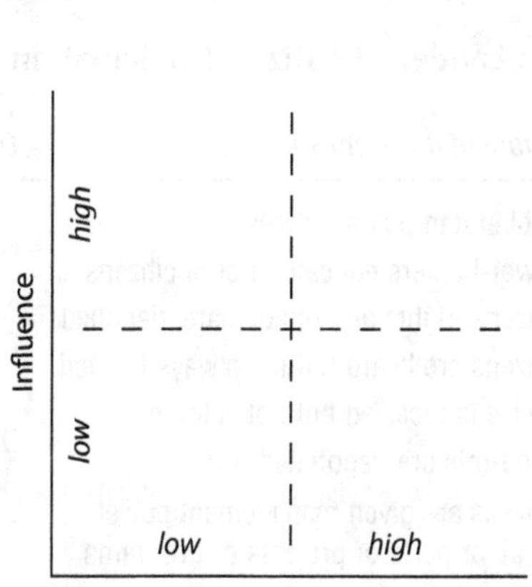

Figure 11.1 Influence versus stake in outcome in identifying publics.

At the other end of the spectrum are publics with a limited stake in the outcome but who are highly influential. This group might include quasi-regulatory bodies, the media, and other special interest groups. Caution must be taken to ensure that the voice of this highly influential group is not overrepresented relative to that of the 'victims'. The final group of publics is those with both high stakes in the outcome and a high degree of influence over the process. For regulators and proponents, this is the most complex group of publics, since they have a potential for considerable gains and losses from project development and at the same time are highly influential over the EIA process and project success. In the Canadian context, some of the more influential Aboriginal groups and quasi-government bodies could be classified as high-stake and high-influence interests.

Levels of Involvement

Public involvement, while an important concern in EIA, rarely consists of highly participatory approaches in which proponents are willing to significantly alter project design or implementation plans. Arnstein (1969), for example, observed that different levels of public involvement could be identified, ranging from manipulation of the public to citizen control (Box 11.2). At one end of the spectrum is 'non-participation'—involvement of the publics in a way that does not include direct participation. This approach consists of what Arnstein labels 'rubberstamp committees' and efforts to inform or 'educate' the public rather than to genuinely seek their involvement. At the opposite end of the spectrum is citizen power in which the affected public is granted full control and authority in development decision-making—rarely, if ever, the case in EIA. In practice, public involvement in EIA has focused primarily on *consulting* or *informing* the public and sometimes negotiating trade-offs, with limited options for greater participation. In other words, public participation in EIA typically involves *providing information* to the public about the proposed project and, at best, degrees of tokenism. Whether this is a sufficient level of public involvement in development impact assessment and decision-making is a matter of debate.

Box 11.2 Arnstein's Ladder of Citizen Participation

Participation	*Nature of involvement*	*Degree of power*
1. Manipulation	Rubberstamp committees	Non-participation
2. Therapy	Power-holders educate or cure citizens	
3. Informing	Citizens' rights and options are identified	Degrees of tokenism
4. Consultation	Citizens are heard but not always heeded	
5. Placation	Advice is received but not acted on	Degrees of citizen power
6. Partnership	Trade-offs are negotiated	
7. Delegated power	Citizens are given management power	
8. Citizen control	for all or parts of projects or programs	

Source: Arnstein 1969.

Provisions and Requirements for Public Participation in Canada

Provisions for public participation in Canadian EIA vary from one jurisdiction to the next. At the federal level, there are five key provisions for public participation. These provisions are also reflected, but vary considerably in their nature and details, under provincial and territorial environmental systems. In many respects, they can be considered the *minimum* provisions for effective and meaningful public participation.

Adequate notice: When a development proposal is submitted to a federal authority and the Canadian Environmental Assessment Act is triggered, a 'notice of assessment' must be posted to the Canadian Environmental Assessment Agency website within 14 days of the start of the assessment.

Access to information: Documents relevant to a project and the assessment process, including the project registration and description, various ministerial decision, the EIS, and often public submissions, are made available to the general public and interested parties through the on-line Canadian Environmental Assessment Registry.

Public comment: Provisions for public comment vary by the type of EIA and are relatively weak for screening level assessments (refer to Chapter 5). Section 18(3) of the Canadian Environmental Assessment Act delegates to the responsible authority the discretion to determine whether public comment is necessary and the nature and extent of that participation. Several criteria are provided in the Canadian Environmental Assessment Agency's 'Public participation guide' to help determine whether public comment in a screening assessment is warranted (Box 11.3). If the responsible authority determines that public comment is necessary, adequate public notice must be given, and the public must have an opportunity to access and comment on the screening report. For a class screening assessment, however, public consultation is required, and the public must be given the opportunity to review and comment on the project before a decision is made to declare the screening a class screening. Only the panel review assessment process guarantees the opportunity for public comment and at various stages of the assessment process; however, about 99 per cent of EIAs completed each year are screening assessments.

Public hearings: Typically, public hearings are associated with review panel assessments and managed by the review panel itself; however, the minister of environment does have the authority to call a public hearing for any assessment. In the case of a review panel assessment, the public must be notified and provided access to documents relating to the environmental assessment. Members of the public may participate in review panels to provide information and appear in public hearings. Public hearings provide a formal opportunity for the public to be heard, for public information and knowledge to be collected, and for public concerns to be formally documented. The information gathered through a hearing process is used to support a review panel's recommendations. Public hearings and review panels are not decision-making bodies.

Participant funding: Also referred to as *intervener funding*, participant funding is usually reserved for larger-scale comprehensive study and review panel assessments.

Box 11.3 Determining Whether Public Participation Is Appropriate for Screening Assessments

Indication of public interest: Is there a known public interest in the project or EIA based on evidence from formal communication, informal communication, or media reports?

History of involvement: Are there affected publics or interest groups in the area that have a history of being involved or have been involved in other, similar projects in the region?

Value conflict: Is there a potential for the project to trigger conflict between environmental, social, and economic values?

Significant adverse environmental effects: Is there a potential that the project will cause significant, adverse environmental effects?

Learning from community or Aboriginal traditional knowledge: Is there an opportunity to improve the project and assessment process based on local and Aboriginal knowledge?

Uncertainty about environmental effects: Is there uncertainty about potential project effects and whether those effects can be properly mitigated?

Equivalent public participation processes: Will the project be subject to other, similar public participation processes that meet the requirements of participation under the EIA process?

Source: Canadian Environmental Assessment Agency's 'Public participation guide'. Available on-line at http://www.ceaa.gc.ca/012/019/index_e.htm.

Funding is available to public groups, organizations, and Aboriginal communities through an application process. Recipients can use the funds to collect their own data or conduct their own research, to participate in project scoping activities and public meetings or hearings, and to support their review of the EIS.

Duty to Consult

> Stakeholder input has generally improved in Canada in the last decade . . . but true meaningful involvement is difficult and has not frequently occurred from a community/First Nations perspective (Lawe, Wells, and Mikisew Cree First Nations Industry Relations Corporation 2005, 207).

There are a variety of provisions and requirements for public participation in EIA; the **duty to consult** is a formal, legal obligation for government to consult with Aboriginal peoples. These legal obligations are separate from and in addition to any requirements for public participation under various federal and provincial environmental assessment systems. In 2004, the Supreme Court of Canada announced two important decisions concerning the *Taku River Tlingit First Nation v. British Columbia* and *Haida Nation v. British Columbia* (Box 11.4). Together, these two decisions would change Aboriginal rights law by specifying the 'duty to consult' with Aboriginal peoples on the part of federal and provincial governments in cases where Aboriginal rights, claims, or titles are known and may be affected by a development or decision but where those rights, claims, or titles have not yet been proven in court. When there is a possibility that Aboriginal or treaty right may be infringed, the government involved has a duty to consult with the affected First Nations group. Such consultation must be carried out with the goal of addressing the concerns of the affected First Nation, but the extent of consultation is determined on a case-by-case basis according to the severity of the potential impact; the extent to which there is an asserted claim or treaty right; the status, merit, or strength of that claim; and whether the Aboriginal or treaty right potentially affected is already established or claimed but not yet established (Potes, Passelac-Ross, and Bankes 2006). The duty to consult applies even before the Aboriginal claim, right, or title is established conclusively by the courts (Brackstone 2002). There is no duty to reach an agreement during consultation; rather, the intent is to substantially address the concerns of the affected Aboriginal peoples. Consent during consultation is not a requirement.

Traditional Ecological Knowledge

The duty to consult should not be confused with the integration of **traditional ecological knowledge** (TEK) into the EIA process. Indeed, the duty to consult can be met in the absence of any consideration of TEK. Often used interchangeably with 'local knowledge' or 'indigenous knowledge', TEK is defined as 'a cumulative knowledge,

Box 11.4 Duty to Consult and the Haida Case Ruling

In 2004, the Supreme Court of Canada released its decisions in *Haida Nation v. British Columbia (Minister of Forests) and Weyerhaeuser.* The Haida case involved a judicial review, pursuant to the British Columbia Judicial Review Procedure Act, of the minister's decision to replace and approve the transfer of a tree farm licence. In 1961, tree farm licences were issued to MacMillan Bloedel, a forest harvesting company, permitting the company to harvest in an area of Haida Gwaii, the Queen Charlotte Islands. The licences were replaced in 1981, 1995, and 2000; the tree farm licence was transferred to Weyerhaeuser in 1999. The Haida challenged these replacements and the transfer, arguing that they were made without the consent of the Haida and over their objections. The Haida's case was dismissed by the British Columbia Supreme Court, which noted that the law could not presume the existence of Aboriginal rights based only on their assertion and concluded that the government had only a 'moral duty' to consult. The decision was appealed, and the court of appeal found that the government had fiduciary obligations of good faith to the Haida with respect to their claims to Aboriginal title and right and that both the province and Weyerhaeuser were aware of the Haida's claims to the area covered by the licence. The court concluded that both the Crown and Weyerhaeuser had a legally enforceable duty to consult with the Haida in an attempt to address their concerns. Weyerhaeuser's appeal of the decision to the Supreme Court of Canada was allowed, but the government's appeal was dismissed. The Supreme Court of Canada ruled that the duty to consult rests with government and government must consult and accommodate when there is knowledge of the potential existence of an Aboriginal right or title, regardless of whether that right or title has been legally established.

Source: Bergner 2005.

practice, and belief, evolving by adaptive processes and handed down through generations by cultural transmission, about the relationship of living beings (including humans) with one another and with their environment' (Berkes 1999, 8). In other words, TEK is associated with societies with historical continuity in resource use in a particular region. At its foundation is knowledge of the land, animals, and the local environment, followed by knowledge of management systems, values, social institutions, and a particular world view. Although TEK is not attached to any particular group in society, it is used largely in an indigenous or Aboriginal context.

TEK and Western science. Perhaps the best way to understand TEK is to contrast it with Western science. According to Peters (2003), TEK is embedded in local culture and communities bounded by the local environment, has a significant moral and ethical context, and emphasizes the absence of separation between nature and culture. Western science, in contrast, is highly analytical, hierarchical in structure, and compartmentalized (e.g., VECs) and authoritative and bureaucratic in nature (Table 11.2). At the same time, however, there are similarities between TEK and Western science knowledge systems. Berkes, Berkes, and Fast (2007), for example, in a study of bowhead whales in the Canadian North, found that TEK was very similar to science in many respects in that many indigenous peoples recognize and monitor various environmental indicators or 'signs' and 'signals' of change.

Table 11.2 TEK and Western Science–Based Knowledge Systems

	TEK	*Western Science*
Basic characteristics	Holistic Experimental Intuitive Subjective	Reductionist Positivist Analytical Objective
Structural governance	Non-hierarchical Egalitarian	Hierarchical Centralized
Communication	Oral record Teaching	Written record Instructive
Decision-making	Consensus	Authoritative
Effectiveness		
Data creation	Slow, inclusive	Fast, selective
Data type	Long time series, single location	Short time series, larger area
Prediction	Short-term cycles	Short-term linear
Explanation	Spiritual and cumulative knowledge	Hypotheses and theory
Biological class	Ecological	Genetic and hierarchical
Experts	Elders Hunters and trappers Communities	Scientists Academics Regulators

Sources: Table compiled by Jasmine Birk, University of Saskatchewan. Based on Berkes, Berkes, and Fast 2007; Houde 2007; Whitye 2006; Peters 2003; Usher 2000; Berkes 1999; Nadasdy 1999; Johannes 1993; Johnson 1992; Wolfe et al. 1992.

Benefits and challenges to TEK integration. The integration of TEK in Canadian EIA is a policy requirement, and various international organizations, such as the United Nations and the World Bank, recognize the value of TEK as an important means of integrating indigenous communities in environmental assessment and decision-making processes. Some believe that incorporating TEK in EIA will help to improve communication between proponents and governments and the local Aboriginal community and create a better, overall understanding of the environment, thus providing an opportunity to improve impact assessment and management (Nadasdy 1999). Others, however, such as Howard and Widdowson (1996), argue that efforts to integrate TEK in EIA may be less than beneficial to decision-makers and to the EIA process in general. Howard and Widdowson suggest that in the case of the BHP Ekati diamond mine in the Northwest Territories, Aboriginal participants 'obfuscated the Panel's attempts to understand the concept' and go on to argue that 'the integration of TEK hinders rather than enhances the ability of governments to more fully understand ecological processes since there is no mechanism, or will, by which spiritually based knowledge claims can be challenged or verified.'

While Howard and Widdowson's view does not represent current thinking on the value of TEK as illustrated by recent literature, it may represent the 'unspoken' view reflected in the current approach to TEK in EIA. Integrating science and TEK is often an exercise in combining two alternative sets of data, but in most instances the science remains relatively unchanged. Houde (2007), on the basis of a review of TEK and co-management boards, observed that '. . . schemes to involve First Nations in decision-making processes have been created for equating TEK to a collection of data about the environment that could complement and be integrated within the existing data sets used by state management systems and for failing to acknowledge the value and cosmological context within which this traditional knowledge was generated and makes sense.'

METHODS AND TECHNIQUES

There are many different methods and techniques for public involvement and communication, and the capability and capacity of these methods and techniques vary considerably. Many different methods and techniques may be used in any single EIA and at each stage of the EIA process, for different publics and for different purposes (Table 11.3). For example, a proponent may hold information seminars early in the project design stage for the communities or interest groups most likely to be affected, whereas other communities or publics outside of the direct project impact zone may be informed at this point only through the mass media. The specific methods and techniques selected for public involvement in any EIA depend on:

- the proponent's objectives;
- the proponent's commitment to public involvement;
- the nature of legal requirements for participation;
- the sensitivity of the receiving environmental and socio-economic environment;
- the availability of time and resources;
- the magnitude of potential project impacts;
- the level of public interest.

Table 11.3 Selected Techniques for Public Involvement and Communication

Capability
meets the criterion = ✓

Capacity
high = ◎
medium = ◉
low = ○

	Capability				*Capacity*		
	provide information	*obtain feedback*	*resolve conflict*	*identify problems and values*	*two-way communication*	*caters to special interests*	*number of people involved*
Public meetings	✓	✓		✓	◉	○	◉
Public displays	✓	✓			◉	○	◎
Presentations to small groups	✓	✓		✓	◉	◉	○
Workshops		✓	✓	✓	◎	◎	○
Advisory committees		✓		✓	◎	◎	○
Public review of EIS	✓	✓		✓	○	◎	◉
Information brochures	✓				○	◉	◉
Press release inviting comments	✓	✓			○	○	◎
Public hearings		✓		✓	○	○	◉
Task forces				✓	◎	◎	○
Site visits	✓			✓	◎	◎	○
Information seminars	✓	✓			◎	◎	○
Mass-media information	✓				○	○	◎

Sources: Based on Sadar 1996; Westman 1985.

MEANINGFUL PARTICIPATION

Canadian ministerial guidelines for public participation under the Canadian Environmental Assessment Act state: 'The public should have an opportunity to have a say in decisions that affect their lives through a meaningful public participation process.' The guidelines recognize that public participation in the EIA process does not necessarily ensure meaningful participation. Based on the International Association for Impact Assessment's (André et al. 2006) international best-practice principles for public participation in EIA, meaningful participation requires adherence to a number of basic and operating principles.

Basic Principles

The basic principles are fundamental to public participation in EIA and apply to all stages of the EIA process and to all tiers, from project-level impact assessment to applications at the broader planning and policy tiers. They are summarized below, based on the IAIA:

Adapted to context: Understanding the context (social institutions, values, culture, history, politics) of the places and communities affected by the proposed development.

Informative and proactive: Early and meaningful provision of information to communities or populations that may be affected by or have an interest in the proposal.

Adaptive and communicative: Recognition of the heterogeneity of affected populations based on differences in values, demographics, knowledge, and interests.

Inclusive and equitable: Ensuring that represented and unrepresented groups and interests are included in participation, including the concerns of future generations.

Educative: Contributing to mutual understanding and respect.

Cooperative: Promoting cooperation, convergence, and consensus-building.

Imputable: Improving the proposal under consideration and reporting back to participants on how their involvement has contributed to decision-making.

Operating Principles

The various operating principles concern how the above basic principles should be applied in the EIA process. They are summarized below, based on the IAIA:

Initiated early and sustained: Participation should commence early in the EIA process, before major decisions are made, and be sustained throughout the process.

Well planned and focused on negotiable issues: Emphasis should be placed on negotiable issues relevant to decision-making, and the objectives, organization, and procedures of participation should be clear to all those involved.

Supportive to participants: There should be adequate information and financial support to facilitate participation, and capacity-building support should be provided to interested groups who lack the capacity to participate.

Tiered and optimized: Participation should occur at the most appropriate level of decision-making.

Open and transparent: Those interested in participating should have access to all relevant information, and that information should be available in an understandable format.

Context-oriented: Methods and procedures for participation should be adapted to the local community, social, cultural, and political context.

KEY TERMS

active publics
duty to consult
inactive publics
traditional ecological knowledge

STUDY QUESTIONS AND EXERCISES

1. Suppose the development of a new waste disposal facility was proposed for your area. Identify who the 'active' and 'inactive' publics might be. Based on 'Arnstein's ladder', at what level would you suggest each of these different groups be involved?
2. Identify a recent controversial development project in your region, and assess how public involvement was conducted. Do you think the outcome of the project proposal or the current situation might have been different if a different degree of public involvement had been incorporated?
3. Discuss the advantages and disadvantages to the proponent of involving publics early in the EIA and project design process.

REFERENCES

André, P., et al. 2006. *Public Participation: International Best Practice Principles.* Special Publication Series no. 4. Fargo, ND: IAIA.

Arnstein, S. 1969. 'A ladder of citizen participation'. *Journal of the American Institute of Planners* 35: 216–24.

Bergner, K. 2005. *The Crown's Duty to Consult and Accommodate.* Paper presented at the Canadian Institute's 5th Annual Advanced Administrative Law and Practice, Ottawa.

Berkes, F. 1999. *Sacred Ecology: Traditional Ecological Knowledge and Resource Management.* Philadelphia: Taylor and Francis.

Berkes, F., M. Berkes, and H. Fast. 2007. 'Collaborative integrated management in Canada's North: The role of local and traditional knowledge and community-based monitoring'. *Coastal Management* 35: 143–62.

Brackstone, P. 2002. *Duty to Consult with First Nations.* Victoria, BC: Environmental Law Centre.

Diduck, A. 2004. 'Incorporating participatory approaches and social learning'. In B. Mitchell, Ed., *Resource and Environmental Management in Canada*, 3rd edn, 497–527. Toronto: Oxford University Press.

Eckel, L., K. Fisher, and G. Russell. 1992. 'Environmental performance measurement'. *CMA Magazine* (March): 16–23.

Houde, N. 2007. 'The six faces of traditional ecological knowledge: Challenges and opportunities for Canadian co-management arrangements'. *Ecology and Society* 12 (2): 34.

Howard, A., and F. Widdowson. 1996. 'Traditional knowledge threatens environmental assessment'. *Policy Options* (November): 34–6.

Johannes, R. 1993. 'Integrating traditional ecological knowledge and management with environmental impact assessment'. In J. Inglis, Ed., *Traditional Ecological Knowledge: Concepts and Cases.* Ottawa: International Program on Traditional Ecological Knowledge, IDRC.

Johnson, M. 1992. 'Research and traditional environmental knowledge: Its development and its role'. In M. Johnson, Ed., *Lore: Capturing Traditional Environmental Knowledge.* Yellowknife, NT: Dene Cultural Institute and IDRC.

Lawe, L.B., J. Wells, and Mikisew Cree First Nations Industry Relations Corporation. 2005. 'Cumulative effects assessment and EIA follow-up: A proposed community-based monitoring program in the oil sands region, northeastern Alberta'. *Impact Assessment and Project Appraisal* 23 (3): 205–9.

Mitchell, B. 2002. *Resource and Environmental Management*. 2nd edn. New York: Prentice Hall.

Murphy, B., and R. Kuhn. 2001. 'Setting the terms of reference in environmental assessments: Canadian nuclear fuel waste management'. *Canadian Public Policy* 27 (3): 249–66.

Nadasdy, P. 1999. 'The politics of TEK: Power and the integration of knowledge'. *Arctic Anthropology* 36 (1–2): 1–18.

Noble, B.F. 2004. 'Integrating strategic environmental assessment with industry planning: A case study of the Pasquai-Porcupine forest management plan, Saskatchewan, Canada'. *Environmental Management* 33 (3): 401–11.

Noble, B., and J. Bronson. 2007. *Models of Strategic Environmental Assessment in Canada*. Ottawa: Canadian Environmental Assessment Agency.

Peters, E. 2003. 'Views of traditional ecological knowledge in co-management bodies in Nunavik, Quebec'. *Polar Record* 39 (208): 49–60.

Petts, J. 1999. 'Public participation and environmental impact assessment'. In J. Petts, Ed., *Handbook of Environmental Impact Assessment*. London: Blackwell Science.

Potes, V., M. Passelac-Ross, and N. Bankes. 2006. 'Oil and gas development and the Crown's duty to consult: A critical analysis of Alberta's consultation policy and practice'. Paper no. 14 of the Alberta Energy Futures project. Calgary: ISEEE

Sadar, H. 1996. *Environmental Impact Assessment*. 2nd edn. Ottawa: Carleton University Press.

Shepherd, A., and C. Bowler. 1997. 'Beyond the requirements: Improving public participation in EIA'. *Journal of Environmental Planning and Management* 40 (6): 725–38.

Usher, P. 2000. 'Traditional ecological knowledge in environmental assessment and management'. *Arctic* 53 (2): 183–93.

Westman, W. 1985. *Ecology, Impact Assessment, and Environmental Planning*. New York: John Wiley.

Whitye, G. 2006. 'Culture in collision: Traditional knowledge and Euro-Canadian governance processes in northern land claim boards'. *Arctic* 59 (4): 401–14.

Wolfe, J., et al. 1992. 'The nature of indigenous knowledge and management systems'. In J. Wolfe et al., Eds., *Indigenous and Western Knowledge and Resource Management Systems*. Guelph, ON: School of Planning and Development, University of Guelph.

PART IV

Advancing Principles and Practices in Environmental Impact Assessment

CHAPTER 12

Cumulative Environmental Effects

THE CONCEPT OF CUMULATIVE EFFECTS

The notion of cumulative environmental change is not new to EIA, and the terms 'cumulative impacts' and 'cumulative effects' actually appeared in many national EIA guidelines and legislations during the early 1970s. The US Council on Environmental Quality (1978), for example, suggested that project impacts on the environment can interact with other past, present, and reasonably foreseeable actions to generate collectively significant environmental change. But even at that time, however, it was observed that EIA had failed to adequately consider:

- the additive effects of several development projects on environmental components;
- the effects of secondary activities resulting from primary development;
- non-linear and indirect environmental responses to development;
- synergistic impacts;
- various impact interactions over time.

It was not until the late 1980s that cumulative effects started to receive any real attention in EIA. In Canada, cumulative effects assessment (CEA) emerged on the scene in the early to mid-1980s as a priority of the Canadian Environmental Assessment Research Council (CEARC). Federally and provincially, CEA is now an accepted part of most project-based EIA frameworks and applications, and in 1995 CEA became mandatory in Canada for all EIAs under the Canadian Environmental Assessment Act. Section 16.1(a) of the Act requires that assessments shall consider:

> . . . the environmental effects of the project, including the environmental effects of malfunctions or accidents that may occur in connection with the project and any cumulative environmental effects that are likely to result from the project in combination with other projects or activities that have been or will be carried out.

The intent was to allow questions of a broader nature related to ecological thresholds and synergistic effects to be assessed in project assessment. In that sense, CEA was viewed as a means of strengthening project EIA and helping it to fulfill its sustainability mandate. Only in recent years, however, have we begun to systematically assess cumulative environmental effects in EIA practice, and there remains considerable room for improvement. In Canada's western prairie watersheds, for example, Schindler and Donahue (2006) suggest an impending water crisis, arguing that the cumulative effects of climate warming, drought, and human activity have seldom, if ever, been considered by policy decision-makers and planners. Rather, the focus of attention has been on project-by-project decision-making, while cumulative environmental change and broader regional and non-point sources of stress have been ignored.

Definition of Cumulative Effects

Simply put, the environmental effects of concern to thinking people are not the environmental effects of a particular development project; they are the cumulative effects of everything (Ross 1994). The terms 'cumulative environmental change', 'cumulative effects', and 'cumulative impacts' are often used interchangeably. Generally speaking, these terms all refer to effects of an additive, interactive, synergistic, or irregular (surprise) nature, caused by individually minor but collectively significant actions that accumulate over space and time (Canter 1999). There is no universally accepted definition of cumulative effects, and various definitions have been proposed in the literature, for example:

- the accumulation of human-induced changes in VECs across space and over time that occur in an additive or interactive manner (Spaling 1997);
- the impact on the environment [that] results from the incremental impact of the action [under review] when added to other past, present, and reasonably foreseeable future actions (US Council on Environmental Quality 1978);
- changes to the environment caused by an action in combination with other past, present, and future actions (Hegmann et al. 1999).

Perhaps the most commonly used definition of 'cumulative environmental effect' is the one provided by the US Council on Environmental Quality (1997), which characterized cumulative environmental effects as:

- the total effect, including direct and indirect, on a given resource, ecosystem, or human community of all actions taken;
- effects that may result from the accumulation of similar effects or the synergistic interaction of different effects;
- effects that may last for many years beyond the life of the action that caused them;
- effects that must be analyzed in terms of the specific resource, ecosystem, or human community affected and not from the perspective of the specific action that may cause them;
- effects that must be approached from the perspective of carrying capacity, thresholds, and total sustainable effects levels.

Sources of Cumulative Effects

The Canadian Environmental Assessment Research Council (1988) defined cumulative effects as occurring when impacts on the biophysical or human environments take place frequently in time or densely in space to such an extent that they cannot be assimilated or when the impacts of one activity combine with the activities of another in a synergistic manner. This suggests that a variety of different *sources* of change contribute to cumulative environmental effects (Table 12.1). Consider, for example, the total downstream effects on water quality and fish resulting from upstream **point source** and **non-point source** stress in a watershed (Figure 12.1), including:

- increased sedimentation due to forestry activity;
- alterations in flow at a hydroelectric facility;
- increased methyl-mercury concentrations caused by reservoir flooding;
- bank erosion at a transmission line crossing;
- water withdrawal and discharge from heavy industry;
- septic leakage from residential areas;
- urban stormwater runoff as a result of surface imperviousness;
- nitrogen loadings from agricultural runoff;
- pharmaceuticals and other chemicals from industry and manufacturing.

The total environmental effect of all of these activities, combined with larger-scale stress caused by climate change and transboundary effects acting on a single VEC, such as fish or water quality, is a cumulative environmental effect. The problem is that not all of these point and non-point sources would be subject to EIA, and certainly few assessment ever would consider non-point sources of stress or capture the point sources from headwater to mouth.

Table 12.1 Sources of Change That Contribute to Cumulative Environmental Effects

Source of change	*Characteristics*	*Example*
Space crowding	High spatial density of activities or effects	Multiple mine sites in a single watershed
Time crowding	Events frequent or repetitive in time	Forest harvesting rates exceeding regeneration and reforestation
Time lags	Activities generating delayed effects	Human exposure to pesticides
Fragmentation	Changes or interruptions in patterns and cycles	Multiple forest access roads cutting across wildlife habitat
Cross-boundary movement	Effects occurring away from the initial source	Acid mine drainage moving downstream to community water supply systems
Compounding	Multiple effects from multiple sources	Heavy metals, chemical contamination, and changes in dissolved oxygen content resulting from multiple riverside industries
Indirect	Second-order effects	Decline in recreational fishery caused by decline in fish populations due to heavy-metal contamination from industry
Triggers and thresholds	Sudden changes or surprises in system behaviour or system structure	Collapse of a fish stock when persistent pressures from harvesting and environmental stress result in a sudden change in population structure

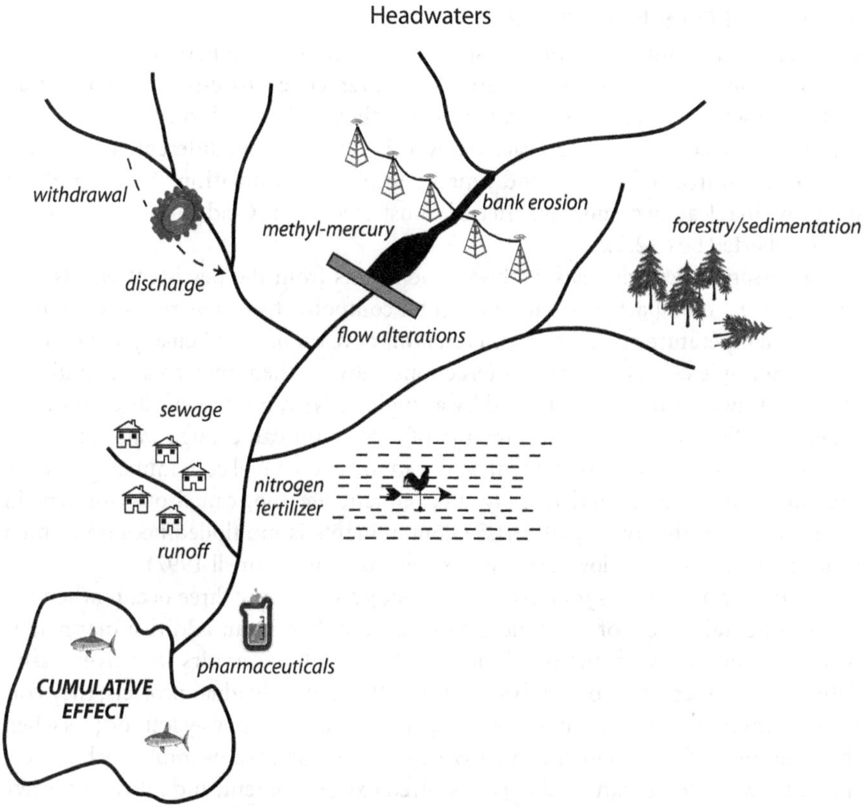

Figure 12.1 Sources of cumulative environmental effects in a watershed.

Types of Cumulative Effects

While multiple types of activities and impacts can lead to cumulative environmental change, Peterson et al. (1987), Sonntag et al. (1987), and Hegmann et al. (1999) identify four broad types of cumulative effects:

1. **Linear additive effects.** Incremental additions to, or deletions from, a fixed storage where each increment or deletion has the same individual effect.
2. **Amplifying or exponential effects.** Incremental additions to, or deletions from, an apparently limitless storage or resource base where each increment or deletion has a larger effect than the one preceding.
3. **Discontinuous effects.** Incremental additions that have no apparent effect until a certain threshold is reached, at which time components change rapidly with very different types of behaviour and responses.
4. **Structural surprises.** Changes that occur as a result of multiple developments or activities in a defined region. They are often the least understood and most difficult to assess.

Pathways of Cumulative Effects

Cumulative environmental effects result from different combinations of actions or pathways that consist of both additive and interactive processes. Peterson et al. (1987) present a classification of functional pathways that lead to cumulative environmental effects (Figure 12.2); each pathway is identified and differentiated according to the sources of change and type of impact accumulation. An example of pathways that lead to cumulative effects is illustrated by the Cold Lake oil sands project in Alberta (Box 12.1).

Single-source perturbations. Pathway one results from the persistent effects of a single project on a particular environmental component, such as repeated changes in water temperature resulting from a reservoir development. When any single activity has multiple effects, potential interactions between them may create cumulative effects. Pathway two is characterized by a single activity, but the effects accumulate synergistically. For example, the creation of a reservoir can change water temperature, lower dissolved oxygen content, and lead to heavy-metal contamination. While each of these effects can individually affect aquatic life, they can also accumulate in such a way that the toxicity of certain contaminants is multiplied because of high water temperatures and low dissolved oxygen content (Bonnell 1997).

Accumulation of effects from two or more projects. Pathway three occurs when the environmental effects of multiple actions accumulate in an additive manner, as would be the case with the development of multiple reservoirs in a river basin. Although no interaction occurs between the effects of individual projects, they collectively result in significant impacts on aquatic resources. Pathway four occurs when these multiple effects do interact in a synergistic manner. For example, each project may alter water temperature, change dissolved oxygen content, and introduce heavy metals, thereby contaminating aquatic life, but the impacts from the interaction of these effects across all projects would be greater than the sum of the individual project impacts (Bonnell 1997).

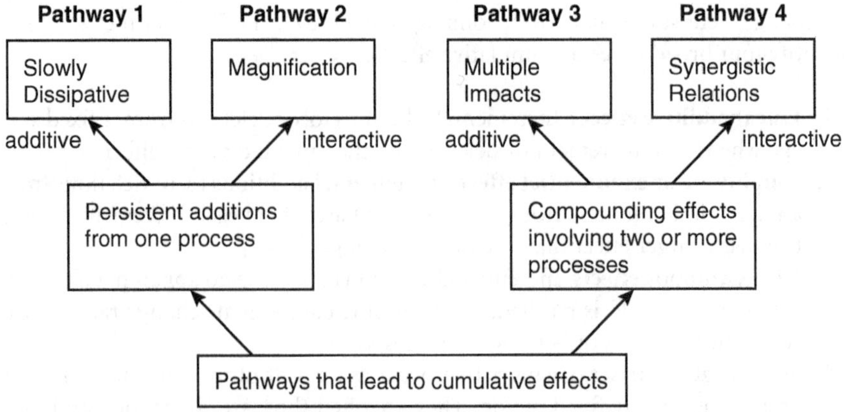

Figure 12.2 Pathways leading to cumulative effects.

Box 12.1 Cold Lake Oil Sands Project Cumulative Effects Pathways

The Cold Lake oil sands project is a heavy oil facility in northern Alberta. Approximately 2,500 wells are currently operating in the region. The Cold Lake facility is currently the second largest producer of oil in Canada. In 2003, the Cold Lake operations accounted for 10 per cent of Canada's crude oil production. Oil deposits are located in sand deposits approximately 400 metres below the surface and are extracted by a steam recovery process that injects high-pressure steam into the reservoir to separate the sand and oil. Wells are drilled and steam injected via clusters of vertical and directional-drilled wells, organized onto large surface pads. In 1997, the proponent, Imperial Oil Resources Limited, proposed to expand its operation in the Cold Lake area with the development of a central plant and additional production wells. A total of 35 impact models were contained in the EIA to assess the cumulative effects of the project on surface water quality, including the additive effects of roads and facilities (well pads) on sediment and contaminant levels in nearby water bodies.

Cumulative impact statement:

- Operation and maintenance of roads and facilities will result in the generation of sediment and transport of contaminants to receiving waters.

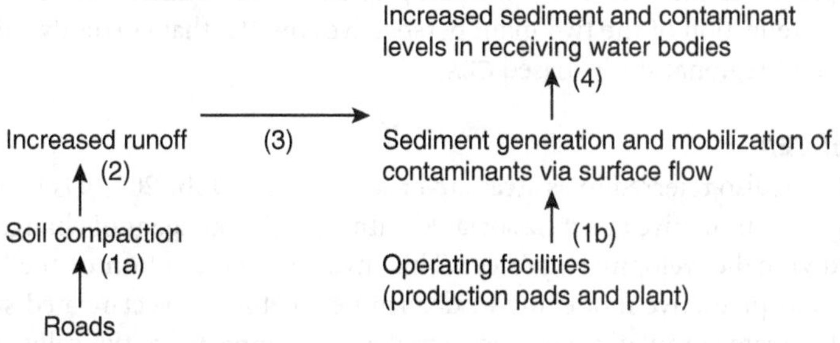

Pathways:

1a. The operation and maintenance of roads will lead to compaction of the roadbed.
1b. Operation and maintenance of pads and plant facilities will result in the generation of sediment and mobilization of contaminants via overland flow from these facilities.
2. Compaction will cause an increase in surface runoff from the road.
3. Increased runoff from roads will result in erosion of exposed soils, resulting in an increase in sediment generation and transport. Soluble contaminants from the road and the roadbed will be transported along with the sediment.
4. Increased sediment and contaminant transport will result in higher levels of these parameters in receiving waters, which will result in a decline in surface water quality.

Sources: Based on Hegmann et al. 1999; Imperial Oil Resources Ltd 1997.

CUMULATIVE EFFECTS ASSESSMENT

Cumulative effects assessment (CEA) refers to the process of systematically analyzing and assessing cumulative environmental change (Spaling and Smit 1994)—that is, identifying environmental effects and pathways in order to avoid, wherever possible, the potential triggers or sources that lead to cumulative environmental change. The main steps in CEA are:

- identifying the affected VECs;
- determining what past, present, and reasonably foreseeable future activities have affected or will affect the VECs and what has led to these activities;
- identifying the potential effects on the VECs of the project or plan in combination with the effects of other activities;
- determining the significance of those effects;
- identifying how to manage cumulative effects (Ross 1998).

It has been said by many that CEA is simply EIA done right. If this is the case, then an argument could be made that EIA has never been done right. Duinker and Greig (2006), for example, conclude that 'the promise and the practice of CEA are so far apart that continuing the kinds and qualities of CEA currently practised in Canada is doing more damage than good.' In practice, CEA is often an 'add-on' to EIA, done after the initial identification and assessment of effects have been completed. This is a poor approach to understanding the complexities of cumulative effects but at the same time a reflection of the two main perspectives on CEA that currently exist: EIA-based CEA and regional study–based CEA.

Models of CEA

EIA-based CEA, also referred to as **stressor-based CEA** (see Dubé 2003), is focused on predicting the cumulative effects associated with a particular agent of change—usually an individual development project. This involves a description of the baseline condition and predictive modelling to determine whether project-related stressors are likely to create cumulative environmental effects. Impacts are typically analyzed on a VEC-by-VEC basis for each individual stressor (Table 12.2). The assumption is that all stressor and stressor–VEC linkages are known within the spatial and temporal bounds of the project's activities (Harriman and Noble 2008). The challenges to this approach are many, including the assumption that the relationship between

Table 12.2 Characteristics of EIA-based CEA

Typical proponent	Single project proponent/developer
Trigger	Project development under an EIA regulation or law
Scope	Inward-focused on project activities and stressors
Temporal bounds	Project life cycle, considering also past environmental change
Spatial bounds	Site-specific, focused on direct on-site and off-site project impacts as defined by the reach of the project's activities
Sources and pathways of effects	Individual, predicted direct project actions
Typical CEA questions	What are the additive or direct impacts of the proposed project activity? What are the key stressors?

Source: Based on Harriman and Noble 2008.

stressor and VEC conditions is known and direct and the assumption that cumulative effects occur within the confines of the project's activities. The nibbling effects of human activities within the ecosystem or even within the immediate project environment, many of which require no formal environmental assessment, go unmanaged. Arguably, understanding such broader regional, nibbling, and indirect cumulative effects and relationships beyond the project's scale and scope is within neither the mandate nor the capabilities of the single project proponent (Box 12.2).

CEA within the context of regional studies, sometimes referred to as **effects-based CEA** (see Dubé 2003), concerns measuring actual environmental responses or VEC conditions rather than the stressors themselves. Recognizing that each development activity or disturbance in a region can represent a high marginal cost to the environment, regional CEA studies expand spatial and temporal boundaries to ambitiously address cumulative impacts resulting from a multiplicity of perturbations over space and time (Cooper and Sheate 2004). Emphasis is on understanding the total effects on a particular VEC from all sources of stress (point, non-point, direct, indirect) and comparing these effects to some reference condition in order to determine an actual measure of cumulative change, irrespective of the number and nature of the impacts causing that change. Under this model, the focus of cumulative effects shifts away from the individual project and its localized stressors

Box 12.2 Constraints of EIA-Based CEA: Terra Nova Offshore Oil Project

With an estimated field reserve of more than 400 million barrels, the Terra Nova oil field, discovered in 1984, is located approximately 350 kilometres east-southeast of St John's, Newfoundland, and 35 kilometres southeast of the Hibernia oil field. In 1996, Petro-Canada, primary operator and 34 per cent shareholder, submitted a development application for approval. The proponents commenced public hearings in April 1997, and in August 1997, after taking into consideration the proponent's impact statement and submissions by government agencies, non-governmental groups, and members of the public, a panel recommended approval of the project. Production of the field began in January 2002.

One contentious issue that emerged during public hearings was the scope of CEA and the responsibility of the proponent. Public submissions to the review process emphasized the cumulative and synergistic concerns in the offshore area, including fishery depletions, oceanographic changes, seabird hunting, climate change, and transportation. Potential future petroleum developments on the Grand Banks, including specific project developments within the region and broader plans to establish a larger offshore industry, were also identified as matters that should be considered in evaluating the project's cumulative impacts. The proponent, in contrast, argued that cumulative impacts should be considered for the Terra Nova project only in terms of specific, planned petroleum projects on the Grand Banks. The panel agreed, suggesting that it is not possible to hold the proponents responsible for future developments beyond their control that might interact with the Terra Nova project to produce cumulative environmental effects. It was suggested during the public review process that an additional comprehensive environmental assessment of all the proposed and potential offshore developments was necessary to include, among other things, the rapidly changing fishery conditions in light of petroleum developments.

to allow for questions of a broader nature related to ecological thresholds and synergistic effects (Table 12.3).

At the Canadian federal level, the Canadian Environmental Assessment Act makes explicit reference to the use of regional studies as a means of supporting project-based assessment and in particular of assisting in the consideration of cumulative environmental effects:

> **S.16(2).** The results of a study of the environmental effects of possible future projects in a region, in which a federal authority participates, outside the scope of this Act, with other jurisdictions referred to in paragraph 12(5)(*a*), (*c*) or (*d*), may be taken into account in conducting an environmental assessment of a project in the region, particularly in considering any cumulative environmental effects that are likely to result from the project in combination with other projects or activities that have been or will be carried out.

According to Spaling et al. (2000), however, rarely under this sort of framework is there authority to implement recommendations or to carry forward CEA findings to specific project-based assessments, and because regional studies are conducted in many different jurisdictions and for different purposes, it is unlikely that a consistent model will be developed or utilized (Grzybowski and Associates 2001). Most regional CEA studies remain 'one-offs', disconnected from development decision-making, and have little influence over EIA outcomes.

Ideally, CEA should involve both approaches, with all aspects of a stressor-based CEA done concurrently with EIA processes, resulting in an assessment process that makes no distinction between the two, and effects-based CEA is then directly tied to project and broader regional effects monitoring.

Spatial Boundaries for CEA

> If we pick the right boundaries, we have a better chance of addressing what's going on in the proper scale (Beanlands and Duinker 1983, 49).

Table 12.3 Characteristics of Regional Study–Based EIA

Typical proponent	Government agency; regional monitoring association; academics
Trigger	Noticed regional decline; preparation for major development plan; land-use study
Scope	Outward-focused on VECs and VEC conditions
Temporal bounds	Past and present, establishing current conditions
Spatial bounds	Regional, ecosystem, or watershed-based
Sources and pathways of effects	Multiple projects and non-project stressors, including incremental, interactive, and potentially synergistic effects, combined with natural change and transboundary effects
Typical CEA questions	What are the current, cumulative VEC conditions? Are the effects of multiple initiatives or actions exceeding thresholds or overloading natural or social carrying capacity?

Source: Based on Harriman and Noble 2008.

Perhaps the most significant area of CEA deviation from conventional EIA is the spatial boundary of assessment. Cumulative effects occur over a large spatial scale and over considerably longer time frames, particularly those 'pathway 4' effects identified in Figure 12.2, which are also extremely difficult to assess. Determining the spatial boundaries for a CEA is critical to its success in effectively managing the cumulative impacts associated with development. Boundaries in CEA delimit the spatial extent of the assessment and thus the environments and VECs that are considered. It is generally acknowledged that in order to assess cumulative effects effectively, the spatial boundaries of the assessment must be extended well beyond the project site. However, if large boundaries are defined, only a superficial assessment may be possible, and uncertainty will increase. Moreover, the incremental addition of a single project may seem less and less significant—only a small drop in a large bucket. If the boundaries are small, a more detailed examination may be feasible, but an understanding of the broad context may be sacrificed. In addition, the incremental impacts of a single project may be exaggerated—a large drop in just a small bucket (Figure 12.3). This, of course, depends on the nature of the project and the impacts. For example, by selecting a relatively small spatial boundary, the impacts of emissions from a proposed smelting operation might seem quite insignificant, especially if emission stacks are relatively high. The choice of spatial boundaries for CEA, then, can be to the proponent's advantage or disadvantage.

Establishing the appropriate boundaries for CEA requires consideration of three types of scale.

Spatial scale refers to the actual geographic extent of the assessment and is typically based on one or both of natural boundaries, such as watersheds, or administrative boundaries, such as townships or landownership. This is usually the most common interpretation of 'scale' in CEA; however, it is certainly not the most functional with regard to the actual analysis of cumulative effects.

Analysis scale is used to examine VECs and impacts across space and is represented by such ideas as data resolution, detail, and granularity. In the Cold Lake oil sands project discussed in Box 12.1, for example, the geographic boundaries for wildlife and vegetation were restricted to local township areas, based on the availability of historical and current information on vegetation composition and wildlife habitat, as well as the extent of available aerial photo coverage. Spatial boundaries were determined on the basis of data availability and desired analysis scale.

Phenomenon scale is perhaps the most important type of scale in CEA, since it refers to the spatial units within which various processes operate or function. Thus, in any single assessment, different spatial boundaries may be appropriate for different cumulative effects and for different VECs. The boundaries selected for cumulative effects on air quality might be quite different from those chosen to assess cumulative effects on soil quality or sedentary versus migratory wildlife.

While there is no absolute method for determining the spatial boundaries for any particular CEA, the CEA literature offers a number of guiding principles to assist the practitioner:

- *Adequate scope.* Boundaries must be large enough to include relationships between the proposed project, other existing projects, and the VECs. This means

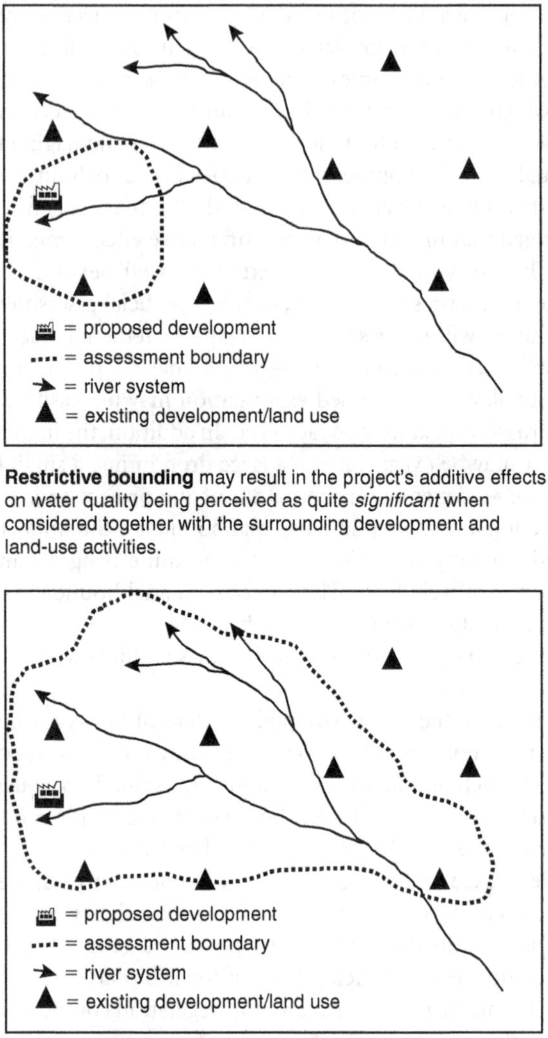

Restrictive bounding may result in the project's additive effects on water quality being perceived as quite *significant* when considered together with the surrounding development and land-use activities.

Ambitious bounding may result in the project's additive effects on water quality being perceived as quite *insignificant* when considered in light of the total effects of all development and land-use activities on the watershed.

Figure 12.3 Restrictive and ambitious spatial bounding.

crossing jurisdictional boundaries if necessary to account for interconnections across systems.

- *Natural boundaries.* Natural boundaries such as watersheds, airsheds, or ecosystems are perhaps the best reflection of the natural components of a system and should be respected.
- *VEC differentiation.* Different VECs and VEC processes operate at different spatial scales, and boundaries must therefore reflect spatial variations in the VECs considered.

- *Maximum zones of detectable influence.* Impacts related to project activities typically decrease with increasing distances; thus, boundaries should be established where impacts are no longer detectable.
- *Multi-scaled approach.* Multiple spatial scales, such as local and regional boundaries, should be assessed to allow for a more in-depth understanding of the scales at which VEC processes and impacts operate.
- *Flexibility.* CEA boundaries must be flexible enough to accommodate changing natural and human-induced environmental conditions.

Generally speaking, however, approaches to cumulative effects assessment seem to be most successful with clearly bounded systems, such as lakes and watersheds, than with more open systems, such as estuaries, marine waters, and terrestrial systems (CEARC 1988).

Temporal Boundaries for CEA

At the heart of cumulative effects assessment is the consideration of the influence of other past, proposed, or likely future activities. Establishing temporal boundaries for CEA requires asking 'how far back in time' and 'how far into the future' should cumulative environmental change resulting from the proposed and other actions and activities be considered in the assessment. The extent of temporal boundaries depends on the amount of information desired, the amount of information available, and what the assessment is trying to accomplish. Examining past conditions may be as simple as examining land-use maps, and in certain cases it may be feasible to incorporate 50 years of historical data if deemed necessary.

The Canadian Environmental Assessment Agency's *Cumulative Effects Assessment Practitioners' Guide* (Hegmann et al. 1999) outlines several options for establishing how far into the past a CEA should extend. The first two options have limited historical perspective and are based on the temporal characteristics of the proposed project itself:

- temporal bounds established only on the basis of existing environmental conditions; or
- when impacts associated with the proposed action first occurred.

Other options are based on more historical perspectives of land use and conditions of environmental change and include:

- the time when a certain land-use designation was made (for example, the establishment of a park or the lease of land for development);
- the time when effects similar to those of concern first occurred; or
- a time in the past representative of desired environmental conditions or pre-disturbance conditions, especially if the assessment includes determining to what degree later actions have affected the environment.

CEA boundaries for *future conditions* are often based on:

- the end of operational life of the proposed project;
- the point of project abandonment and site reclamation; or
- a time when VECs are likely to be restored, considering natural variations, to their to pre-disturbance conditions.

Identifying which potential future actions and activities to include in CEA can be much more uncertain. Hegmann et al. (1999) thus characterize future actions of three types:

1. *Certain actions.* The action will proceed, or there is a high probability that the action will proceed. This includes actions or projects already approved or submitted for approval or that have been proposed by the proponent.
2. *Reasonably foreseeable actions.* The action may proceed, but there is some uncertainty about that conclusion. This might include projects under review for which approval is likely to be conditional, activities identified in an approved or proposed development plan, or induced activities that may occur should the project proposed be approved.
3. *Hypothetical actions.* There is considerable uncertainty as to whether the action will ever proceed. Such actions or activities include those discussed only on a conceptual basis or those speculated to proceed, based on current information. **Scenario analysis** is a common tool used to identify such hypothetical actions.

These actions lie on a continuum from most likely to least likely to occur. For each assessment, the practitioner or the regulatory agency will have to decide how far into the future the assessment should reach. Often, a major criterion is whether the future action or actions are likely to affect the same VECs as the proposal under consideration. While practical, this criterion may detract from these projects, creating 'nibbling' effects that, while they may not directly affect the same VECs, contribute to overall decline in environmental quality.

Methods for CEA

Many of the methods used in EIA can be adapted to address cumulative environmental effects (Table 12.4). The difficulties inherent in each method still exist, however, and in fact seem to be exacerbated by the complexity associated with the nature of cumulative effects. For example, the subjectivity of weighting values in matrix techniques becomes even more problematic when multiple projects are being considered, and the coarseness of data and its analysis often increase as the spatial scale of the assessment increases. Two methods discussed in Chapter 3 are reiterated here within the context of simple additive CEA.

Interaction matrices. Interaction matrices can be used to identify the effects of multiple projects. This requires the subjective assignment of magnitude and significance values to the impacts expected to result from each individual project or from different stages of the same project and 'summing up' the overall cumulative impacts. Figure 12.4 shows a series of impact assessment matrices for both the cumulative effects of a single activity and the cumulative effects of multiple activities on selected VECs. Each matrix is first evaluated individually, based on its magnitude and significance, and is then overlain or summed with each other matrix to generate an overall perspective of cumulative environmental effects.

Weighting methods. A common approach to CEA for simple additive impacts involves weighting methods and techniques. The Cluster Impact Assessment Procedure (CIAP), for example, was developed by the US Federal Energy Regulatory Commission for the CEA of small-scale hydroelectric power projects. At the heart of

Table 12.4 Selected Methods That Support Cumulative Effects Assessment

Method	*Characteristics*
Geographic Information Systems	Provide spatial analysis of cumulative effects through digital mapping. Useful for mapping sources of cumulative change and effects and areas of overlapping project impacts. Limited application for analysis of pathways and restricted by data availability.
Network analysis	A qualitative network based on feedback relationships. Useful for identification of second-, third-, and higher-order effects pathways and relationships but largely untested in CEA practice.
Interactive matrices	Total cumulative impact is assumed to be the sum of project-specific effects derived from overlain or otherwise manipulated impact identification and assessment matrices. Allows the consideration of cumulative effects of multiple sources, but effects are not differentiated by type. Values rely heavily on expert judgment.
Ecological modelling	Computer-based, dynamic modelling of ecosystem components and interactions. Theoretically sound, can accommodate a large volume of data, and easily provides opportunity for scenario analysis of future conditions. Application, however, is dependent on data availability, model validation, and verification of relationships between VECs and is limited to systems for which behaviour is well understood.
Expert weighting and scoring	Assignment of impact scores and impact values to affected VECs to derive a total cumulative impact score. The approach is not limited by data availability and provides a simple, sound basis on which to evaluate the additive impacts of proposed developments across VECs. However, the impact scores and weights are only as informative as the individual who assigns them.

Source: Based on Canter 1999.

the CIAP is the calculation of a 'total impact' score for each of the affected VECs based on individual project impacts and related interaction impacts. Numerical impact ratings are assigned as impact scores, and the relative importance of the components is considered (Canter 1999). A similar approach, using a Delphi process, was adopted by Bonnell (1997) to assess the cumulative environmental effects of multiple existing and proposed small-scale hydroelectric development projects in Newfoundland. Box 12.3 gives an example of weighting impacts and VECs for simple additive effects.

CHALLENGES TO CEA IN CANADA

Notwithstanding the recognized need for CEA and the implications of not doing it, there are constant and consistent messages that CEA is either not being done or not being done well when it is done. These issues and challenges to CEA have only been addressed head-on recently, in a paper by Duinker and Greig (2006), 'The impotence of cumulative effects assessment in Canada: Ailments and ideas for redeployment'. On the basis of observations from practice and experience, Duinker and Greig offer six reasons for the current state of CEA practice in Canada.

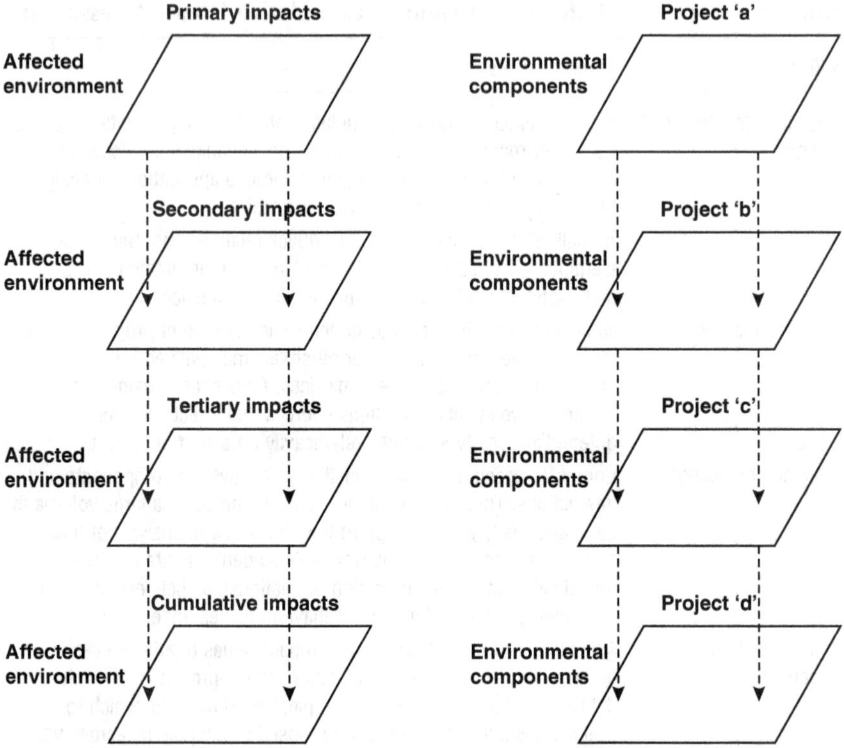

Interaction matrices for a single project generate a cumulative matrix depicting the total primary, secondary, and tertiary effects on particular environmental components. For example, clearing of land may increase erosion (primary impact), which in turn increases runoff (secondary impact), leading to increased stream turbidity (tertiary impact). Similar project activities may similarly generate effects on water quality, contributing to overall cumulative impact.

In a similar fashion, matrices can be overlain to sum the impacts of multiple projects on various environmental components. Although not depicted here, each environmental component can be weighted according to its relative importance, with impact magnitude and significance values incorporated into each project matrix. This approach is useful for proving an overall view of impacts associated with activities in a particular geographic area.

Figure 12.4 Matrix interactions for multiple components of the same project and multiple projects.

Box 12.3 Weighting Impacts and VECs for Simple Additive CEA

Projects: small-scale hydroelectric project '**A**'
small-scale hydroelectric project '**B**'
small-scale hydroelectric project '**C**'

VECs: forest resources, fish resources, caribou

Impact of project 'n' on VEC 'i' $I_n VEC_i = P_i(M_i \times SE_i \times TD_i) \times CS_i$

P_i = the probability that VEC 'i' will be affected by project 'n'

- expressed as % (i.e., likelihood)

continued

M_i = the magnitude of the effect of project 'n' on VEC 'i'

- 1 = negligible: a change to the VEC that is indistinguishable from natural variation.
- 2 = minor: a reversible change to the VEC's normal baseline condition; the fundamental integrity of the VEC is not affected.
- 3 = moderate: a reversible change to the VEC's normal or baseline condition, with a medium probability of second-order effects on other environmental components; the fundamental integrity of the VEC is not threatened.
- 4 = major: an irreversible change to the VEC's normal or baseline conditions, with a high probability of second-order effects on other environmental components; the fundamental integrity of the VEC is threatened.

SE_i = the spatial extent of the effect of project 'n' on VEC 'i'

- 1 = site-specific: effect will be confined to the project development area.
- 2 = local: effect will be confined to the project area and immediate environment.
- 3 = regional: effect will occur within and beyond the development area and immediate environment, affecting a defined territory surrounding the development.
- 4 = ecosystem: effect will occur throughout the ecosystem.

TD_i = the temporal duration of the effect of project 'n' on VEC 'i'

- 1 = short-term: effect may persist less than two years from the onset of disturbance.
- 2 = medium-term: effect may persist from two to less than five years from the onset of disturbance.
- 3 = long-term: effect may persist from five to less than 10 years from the onset of disturbance.
- 4 = prolonged: effect may persist 10 years or more from the onset of disturbance.

CS_i = the current state of VEC 'i' in the vicinity of project 'n'

- 1 = resilient: VEC is quite resilient to impact because of its natural conditions and/or lack of other adjacent human activities.
- 2 = low sensitivity: VEC has a low susceptibility to impact because of its natural conditions and/or the impacts of other adjacent human activities.
- 3 = medium sensitivity: VEC is moderately susceptible to impact because of its natural conditions and/or the impacts of other adjacent human activities.
- 4 = high sensitivity: VEC is highly susceptible to impact because of its natural conditions and/or the impacts of other adjacent human activities.

VEC Imp 'i' = the relative importance of VEC 'i' compared to all other VECs with regard to (for example):

a) socio-economic dependency
b) human health
c) ecological functioning

Step 1: Determine relative importance (significance) of affected VECs based on:

a) socio-economic dependency

Using Saaty's paired comparison methods, demonstrated in Chapter 8, the following matrices present a hypothetical example of VEC weighting. For example, in terms of socio-economic dependency, 'forest resources' (row 1) are considered slightly less important than 'fish resources' (column 2) ('1/3') and significantly less important than 'caribou' (column 3) ('1/5').

continued

	Forest resources	*Fish resources*	*Caribou*
Forest resources	1	1/3	1/5
Fish resources	3	1	1/3
Caribou	5	3	1

Divide each cell entry of the paired comparison matrix by the sum of its corresponding column to normalize the matrix:

	Forest resources	*Fish resources*	*Caribou*
Forest resources	0.11	0.08	0.13
Fish resources	0.33	0.23	0.22
Caribou	0.55	0.69	0.65

Determine the priority vector of the matrix by averaging the row entries in the normalized assessment matrix:

- Forest resources = 0.12; Fish resources = 0.26; Caribou = 0.63

Multiply each column of the initial assessment matrix by its relative priority vector, and sum the results:

$$0.12 \begin{bmatrix} 1 \\ 3 \\ 5 \end{bmatrix} + 0.26 \begin{bmatrix} 1/3 \\ 1 \\ 3 \end{bmatrix} + 0.63 \begin{bmatrix} 1/5 \\ 1/3 \\ 1 \end{bmatrix} = \begin{bmatrix} 0.34 \\ 0.83 \\ 2.01 \end{bmatrix}$$

Divide the results by the original priorities to determine the VEC importance based on socio-economic dependency:

- Forest resources = 0.34 / 0.12 = 2.83
- Fish resources = 0.83 / 0.26 = 3.19
- Caribou = 2.01 / 0.63 = 3.19

Follow the same procedure for b) human health and c) ecological functioning to determine overall VEC importance. For example:

	Socio-economic dependency	*Human health*	*Ecological functioning*	*Total VEC importance**
Forest resources	2.83	1.83	1.45	6.11
Fish resources	3.19	2.15	1.24	6.50
Caribou	3.19	3.24	1.89	8.32

*The assumption is that 'socio-economic dependency', 'human health', and 'ecological functioning' are equally important. Otherwise, these factors must be weighted as well and the weight multiplied by each of the respective VEC weights before adding the 'total VEC importance'.

Step 2: Determine the impact of each project on each VEC by $I_nVEC_i = P_i(M_i \times SE_i \times TD_i) \times CS_i$

For example, for project 'A':

continued

	P_i	M_i	SE_i	TD_i	CS_i	I'A'VEC
Forest resources	0.45	3	1	2	3	8.1
Fish resources	0.10	4	3	3	4	14.4
Caribou	0.78	2	2	2	3	18.72

Follow the same procedure for projects 'B' (**I'B'VEC**) and 'C' (**I'C'VEC**), and transfer all values to the total cumulative effects score table shown below.

Step 3: Determine the weighted and cumulative impacts

Project 'A' impact on VEC 'i' × Imp VEC 'i'

	I'A'VEC 3 total VEC importance	*I'B'VEC 3 total VEC importance*	*I'C'VEC 3 total VEC importance*	*Cumulative (additive) impact score*
Forest resources	(8.1)(6.11) = 49.49	(2.21)(6.11) = 13.50	(5.21)(6.11) = 31.83	Sum = 94.82
Fish resources	(14.4)(6.50) = 93.60	(3.24)(6.50) = 21.06	(4.21)(6.50) = 27.36	Sum =142.04
Caribou	(18.72)(8.32) = 155.75	(7.12)(8.32) = 59.23	(3.12)(8.32) = 25.95	Sum = 240.93
Cumulative impact of project 'n' on all VECs =	298.84	93.78	85.14	Overall cumulative impact of all projects on all VECs Sum = 477.78

This approach allows quick insight into the individual contributions of each project to the overall cumulative effects. Project 'A', for example, is contributing proportionately more to overall cumulative impacts when compared to projects 'B' and 'C'. Moreover, it is generating the most significant effects on caribou, one of the VECs deemed to be of greatest overall importance to socio-economic dependency, health, and ecological functioning. While this approach is highly subjective and perhaps no more accurate than the individual(s) assigning the values, it does allow for a systematic examination of the additive effects of individual projects on VECs and establishes a baseline from which effects-based CEA can be undertaken. The approach is also useful in situations where baseline data are not available or to analyze the judgments of experts, publics, or other project interests.

Sources: Based on Irving and Bain 1993; Bonnell 1997; Canter 1999; Noble 2002.

The first reason concerns the context of CEA as currently required in Canada—situated within project-based EIA. As explained at the outset of this chapter, cumulative environmental effects concern the total effects of human activities on a VEC. Project EIA, in contrast, is concerned about project-induced stress and making sure that the impacts of a project are acceptably small rather than understanding the total effects of all stressors, project and non-project, on any single VEC. Second, and closely related, is that project EIA is concerned primarily with minimizing project stress to a level of acceptability. The objective of proponents is to ensure that their project meets regulatory and public approval—this usually means minimizing any efforts regarding CEA and paying little attention to understanding VEC quality and longer-term sustainability.

A third explanation concerns thresholds. As noted in Table 12.3, understanding the cumulative effects of human activity on a VEC, and the implications of such effects, requires some understanding of thresholds and carrying capacities. The challenge, however, is that thresholds are not easily determined, and that is particularly so within the spatial and temporal confines of a project assessment. When thresholds are addressed in EIA, they are usually defined within the context of the project as opposed to the total effects on a VEC and, further, typically defined on the basis of public acceptability as opposed to ecological knowledge.

Fourth, CEA is rarely an integrated part of the EIA process. The practice is to address project-based effects first, identify mitigation options, and then 'add on' a CEA as a separate part of the impact analysis. The assumption is that cumulative effects are a special class of environmental effects and that somehow, most project effects are non-cumulative in nature. This thinking is closely tied to Duinker and Greig's fifth explanation concerning the current state of CEA—that there still is limited understanding of 'what is' a cumulative effect. Finally, CEA and management are ultimately about the future and demand looking far enough into the future to capture the full array of human activities and natural changes that may affect the sustainability of VECs of concern. This is a highly uncertain environment and one that is about possible futures and outcomes—a view that stands in sharp contrast to the shorter-term perspective of project approval and predicting the 'most likely', versus the most desirable, effects of development.

This is not to say that project-based CEAs are not useful; rather, something more is needed to address and manage cumulative environmental change in an effective manner (Cooper 2003; Creasey 2002). There is now general agreement that CEA should go beyond the evaluation of site-specific direct and indirect project impacts to address broader regional environmental impacts and concerns. Cocklin, Parker, and Hay (1992) identify three main objectives in advancing CEA beyond EIA:

- to develop a broader understanding of the current state of the environment vis-à-vis cumulative change processes;
- to identify, insofar as possible, the extent to which cumulative effects in the past have conditioned the existing environment;
- to consider priorities for future environmental management with respect to general policy objectives and with regard to potential development options.

The underlying notion is that cumulative environmental change is the product of multiple, interacting development actions and that the multiplicity of development decisions in a particular region, while often individually insignificant, cumulatively lead to significant environmental change. More than 15 years since Cocklin, Parker, and Hay's observations, however, little progress seems to have been made; CEA is perhaps more entrenched in project-based EIA today than it has been in its more than 25-year history.

KEY TERMS

ambitious bounding
amplifying effects
analysis scale
discontinuous effects
effects-based CEA
linear additive effects
non-point source stress
phenomenon scale
point source stress
restrictive bounding
scenario analysis
spatial scale
stressor-based CEA
structural surprises

STUDY QUESTIONS AND EXERCISES

1. Do provisions exist under your national or provincial EIA system for cumulative effects assessment?
2. Using the example of multiple reservoir developments in a single watershed, sketch a diagram similar to Figure 12.2, and identify and classify the different types of cumulative impacts that might result. State the impact 'pathways' as illustrated by the example in Box 12.1.
3. What is the difference between effects-based and stressor-based approaches to cumulative effects assessment?
4. Using an example, explain how a proponent might use spatial bounding to its advantage. Given this, should the proponent be solely responsible for determining the spatial boundaries for cumulative effects assessment?
5. It has been said that cumulative effects assessment is simply EIA done right. Do you agree? Given the challenges and constraints of EIA in assessing and understanding cumulative effects, should CEA be part of EIA or a separate, independent process? Identify the benefits and limitations of a more integrated versus a more separated CEA process.
6. Cumulative effects often result from multiple and often unrelated project developments in a single region. Should regional cumulative effects assessment be the responsibility of the project proponent?
7. When a project creates environmental damage, it is often the responsibility of the proponent to rectify or compensate for such damage. Assume a region where there are multiple projects and activities, including oil and gas, forestry, highway, recreation, and hydroelectric developments. Individually, each project was approved for development based on the fact that it would not generate significant environmental effects. Cumulatively, however, all of these activities are contributing to overall environmental decline.
   a) Who should be responsible for managing overall cumulative environmental change resulting from the many, unrelated project developments and activities?
   b) How does one determine how much each development or activity is contributing to cumulative change?
   c) Given that each project is 'individually insignificant' but that together they are cumulatively damaging, should an additional development be permitted in the region if it too is determined to be individually insignificant? What are the implications of such a decision with regard to equity versus environmental protection?
8. Assume that three industrial developments are proposed for a previously undeveloped region adjacent to a river. The projects are similar in size, and each will require clearing of a forested site. Several VECs and their overall importance weightings have been identified for the region, including wildlife (2.21), aquatic resources (3.41), water quality (5.73), and air quality (4.68). Using the equation '$I_nVEC_i = P_i(M_i \times SE_i \times TD_i) \times CS_i$',

construct a hypothetical cumulative impact assessment matrix for each project. Assign hypothetical project impact scores, and calculate the following:
a) the cumulative impact of each individual project across all VECs;
b) the cumulative impact of all projects on each VEC;
c) the cumulative impact of all projects across all VECs.
Discuss the advantages and limitations of the above approach to cumulative effects assessment.

REFERENCES

Beanlands, G.E., and P.N. Duinker. 1983. 'Lessons from a decade of offshore environmental impact assessment'. *Ocean Management* 9 (3/4): 157–75.

Bonnell, S. 1997. 'The cumulative effects of proposed small-scale hydroelectric developments in Newfoundland, Canada'. MA thesis, Memorial University of Newfoundland.

Canter, L. 1999. 'Cumulative effects assessment'. In J. Petts, Ed., *Handbook of Environmental Impact Assessment*, v. 1, *Environmental Impact Assessment: Process, Methods and Potential*. London: Blackwell Science.

CEARC (Canadian Environmental Assessment Research Council). 1988. *The Assessment of Cumulative Effects: A Research Prospectus*. Ottawa: Supply and Services Canada.

Cocklin, C., S. Parker, and J. Hay. 1992. 'Notes on cumulative environmental change I: Concepts and issues'. *Journal of Environmental Management* 35: 51–67.

Cooper, L. 2003. *Draft Guidance on Cumulative Effects Assessment of Plans*. EPMG Occasional Paper 03/LMC/CEA. London: Imperial College.

Cooper, L., and W. Sheate. 2004. 'Integrating cumulative effects assessment into UK strategic planning: Implications of the European Union SEA Directive'. *Impact Assessment and Project Appraisal* 22 (1): 5–16.

Creasey, R. 2002. 'Moving from project-based cumulative effects assessment to regional environmental management'. In A.J. Kennedy, Ed., *Cumulative Environmental Effects Management: Tools and Approaches*. Calgary: Alberta Society of Professional Biologists.

Dubé, M. 2003. 'Cumulative effects assessment in Canada: A regional framework for aquatic ecosystems'. *Environmental Impact Assessment Review* 23: 723–45.

Duinker P., and L. Greig. 2006. 'The impotence of cumulative effects assessment in Canada: Ailments and ideas for redeployment'. *Environmental Management* 37 (2): 153–61.

Grzybowski and Associates. 2001. 'Regional environmental effects assessment and strategic land use planning in British Columbia'. Report prepared for the Canadian Environmental Assessment Agency Research and Development Program. Hull, QC: Canadian Environmental Assessment Agency.

Harriman, J., and B. Noble. 2008. 'Characterizing project and regional approaches to cumulative effects assessment in Canada'. *Journal of Environmental Assessment Policy and Management* 10 (1): 25–50.

Hegmann, G., et al. 1999. *Cumulative Effects Assessment Practitioners' Guide*. Prepared by AXYS Environmental Consulting and CEA Working Group for the Canadian Environmental Assessment Agency, Hull, QC.

Imperial Oil Resources Limited. 1997. *Cold Lake Expansion Project Environmental Impact Statement*. Calgary: Imperial Oil Resources Limited.

Irving, J.S., and M.B. Bain. 1993. 'Assessing cumulative impact on fish and wildlife in the Salmon River Basin, Idaho'. In S.G. Hildebrand and J.B. Cannon, Eds, *Environmental Analysis: The NEPA Experience*. Boca Raton, FL: Lewis Publishers.

Noble, B.F. 2002. 'Strategic environmental assessment of Canadian energy policy'. *Impact Assessment and Project Appraisal* 20 (3): 177–88.

Peterson, E., et al. 1987. *Cumulative Effects Assessment in Canada*. Ottawa: Supply and Services Canada.

Ross, W. 1994. 'Assessing cumulative environmental effects: Both impossible and essential'. In A.J. Kennedy, Ed., *Cumulative Effects Assessment in Canada: From Concept to Practice*. Papers from the 15th Symposium held by the Alberta Society of Professional Biologists, Calgary.

———. 1998. 'Cumulative effects assessment: Learning from Canadian case studies'. *Impact Assessment and Project Appraisal* 16 (4): 267–76.

Schindler, D., and W. Donahue. 2006. 'An impending water crisis in Canada's western Prairie provinces'. *Proceedings of the National Academy of Science of the United States* 103: 7,210–16.

Sonntag, N., et al. 1987. *Cumulative Effects Assessment: A Context for Further Research and Development*. Ottawa: Supply and Services Canada.

Spaling, H. 1997. 'Cumulative impacts and EIA: Concepts and approaches'. *EIA Newsletter* (University of Manchester) v. 14. www.art.man.ac.uk/EIA/publications/ index.htm.

Spaling, H., et al. 2000. 'Managing regional cumulative effects of oil sands development in Alberta, Canada'. *Journal of Environmental Assessment Policy and Management* 2 (4): 501–28.

Spaling, H., and B. Smit. 1994. 'Classification and evaluation of methods for cumulative effects assessment'. In A.J. Kennedy, Ed., *Cumulative Effects Assessment in Canada: From Concept to Practice*, 47–65. Papers from the 15th Symposium held by the Alberta Society of Professional Biologists, Calgary.

US Council on Environmental Quality. 1978, 1997. *Considering Cumulative Effects under the National Environmental Policy Act*. Washington: Council on Environmental Quality, Executive Office of the President.

CHAPTER 13

Strategic Environmental Assessment

THE BASIS OF STRATEGIC ENVIRONMENTAL ASSESSMENT

Strategic environmental assessment (SEA) has become one of the most widely discussed issues in the field of contemporary environmental assessment. SEA is essentially about integrating *environment* into higher-order decision-making processes, thereby providing an early, overall analysis of the relationships between **policies**, **plans**, and **programs** (PPPs) and the potential effects of the projects that emerge from those PPPs (Sadler 1996). It is difficult to cover all aspects of SEA in a single chapter, since SEA itself deserves its own separate book. Attention here is limited to selected SEA principles and practices: basic SEA characteristics, the international status of SEA, and illustration of a practical framework for application.

SEA is based on the notion that many of the decisions that affect the environment are made long before project developments are proposed. Moreover, such decisions often affect the nature and type of project proposals. For example, decisions concerning a regional land-use management plan are likely to shape the future development of the region and the specific types of project activities undertaken. **Strategic environmental assessment**, while variously defined, broadly refers to the environmental assessment of PPPs and their alternatives. A strategic approach offers a foundation upon which to base environmental decision-making and ensures the full consideration of alternative options at an early stage when there is greater flexibility with respect to decisions. Thus, SEA is proactive, asking 'what is the preferred option?' and 'what is the preferred attainable end(s)?' rather than predicting the most likely outcomes of a predetermined type of action.

In essence, SEA extends EIA upstream—but at the same time it adopts a different set of principles and criteria from those used in project-based assessment (Box 13.1). Guiding principles and criteria for SEA encompass three components, capturing the entirety of SEA from its institutional foundation to its ultimate influence on decision-making. System components refer to the basic provisions and requirements for SEA and the position of SEA in the broader planning and decision-making environment. Provisions for SEA and an understanding of its role in and contributions to planning and decision-making processes are prerequisites to an effective SEA process. The procedural components concern the various methodological and process elements of SEA—in other words, the practice of SEA. Result components refer to the overall influence of SEA on decision-making and subsequent project-based EIA, including the opportunity for broader system-wide SEA learning and process development as well.

Box 13.1 Principles and Criteria for Strategic Environmental Assessment

System components	*Definition*
1. Provisions	• clear provisions, standards, or requirements to undertake the SEA
2. Integration	• application early enough to address deliberation on purposes and alternatives or to guide initial conception of review for an existing PPP
3. Tiering	• assessment undertaken within a tiered system of environmental assessment, planning, and decision-making
4. Sustainable development	• sustainability/sustainable development a guiding principle and integral concept
Process components	*Definition*
5. Responsibility and accountability	• clear delineation of assessment roles and responsibilities • mechanisms to ensure impartiality/independence of assessment review • opportunity for appeal of process or decision output
6. Purpose and objectives	• assessment purpose and objectives clearly stated • centred on a commitment to sustainable development principles
7. Scoping	• opportunity to develop and apply more or less onerous streams of assessment sensitive to the context and issue • consideration of related strategic initiatives • identification and narrowing of possible valued ecosystem components to focus on those of most importance based on the assessment context
8. Alternatives	• comparative evaluation of potentially reasonable alternatives or scenarios
9. Impact evaluation	• identification of potential impacts or outcomes resulting from each option or scenario under consideration • integration or review of sustainability criteria specified for the particular case and context
10. Cumulative effects	• consideration of potential cumulative effects and life-cycle issues
11. Monitoring program	• procedures to support monitoring and follow-up of process outcomes and decisions for corrective action
12. Participation and transparency	• opportunity for meaningful participation and deliberations • transparency and accountability in assessment process
Result components	*Definition*
13. Decision-making	• identification of a 'best' option or strategic action • authoritative decisions, position of the authority of the guidance provided
14. PPP and project influence	• defined linkage with assessment and review or approval of any anticipated lower-tier initiatives • demonstrated PPP influence, modification, or downstream initiative • identification of indicators or objectives for related or subsequent strategic initiatives or activities
15. System-wide learning	• opportunity for learning and system improvement through regular system or framework review

Source: Based on Noble 2009.

SYSTEMS OF SEA

The origins of SEA are often tied to the US NEPA of 1969. However, the term 'SEA' was first coined in 1989 by Wood and Djeddour (see Partidário 2007), who suggested:

> The environmental assessments appropriate to policies, plans, and programmes are of a more strategic nature than those applicable to individual projects and are likely to differ from them in several important respects. . . . We have adopted the term 'strategic environmental assessment' (SEA) to describe this type of assessment (Wood and Djeddour 1989).

It was not until after a number of high-profile international developments—namely, the World Bank's (1999) recommendation for environmental assessment of policy, the report of the World Commission on Environment and Development, *Our Common Future* (1987), and the United Nations Earth Summit held in 1992 in Rio de Janeiro—that SEA gained international recognition. It is only within the past decade that SEA has been clearly evident in practice.

SEA remains far less advanced than EIA, and only a few nations have formal provisions for SEA systems. Where provisions do exist, they vary considerably (Box 13.2). In the US, for example, provisions for SEA fall under NEPA, where SEA is broadly interpreted to be **programmatic environmental assessment** or area-wide EIA. Hundreds of programmatic assessments are completed in the US each year, which essentially involve direct application of EIA to programs of development.

In the UK, SEA was initially carried out through a less formalized policy and plan environmental appraisal process. Formal requirements for SEA were adopted in the UK in 2004 under European SEA Directive 2001/42/EC. In the Czech Republic, SEA was introduced by means of a reform to formal EIA legislation, whereas in Australia, SEA was at first adopted informally as part of resource management programs and

Box 13.2 Models of SEA Systems

EIA-based: SEA is implemented under EIA legislation, such as in the Netherlands, or carried out under separately administrated procedures, such as in Canada and Hong Kong.

Environmental appraisal: SEA provision is made through a less formalized process of policy and plan appraisal, as in the UK prior to the European SEA Directive.

Dual-track system: SEA is differentiated from EIA and implemented as a separate process, such as the Netherlands' 'Environmental test' or E-test of legislation and plans.

Integrated policy and planning: SEA is an integrated component of policy and plan development and decision-making, such as in New Zealand or for forest management plans in Saskatchewan.

Sustainability appraisal: SEA elements as separate evaluation and decision tools are replaced by integrated environmental, social, and economic assessment and appraisal of policy and planning issues, such as the former Australian Resource Assessment Commission and current sustainability plans in the UK.

Source: Based on UNEP 2002.

then in 1999 was introduced through formal legislated requirements under the broad umbrella of environmental assessment that includes planning strategies and programs (Noble 2003).

Canada is recognized as a country that has made significant contributions to the development of environmental assessment above the project level (Table 13.1). The underlying principle of SEA in Canada is that in order to make informed decisions in support of sustainable development, decision-makers at all levels must integrate social, economic, and environmental considerations with PPP development (see Noble 2009 for an overview of SEA development in Canada and the Canadian Environmental Assessment Agency website www.ceaa.gc.ca for the cabinet directive on SEA).

SEA was formally established in Canada in 1990 by means of a federal cabinet directive and as a separate process from EIA, 'making it the first of the new generation of SEA systems that evolved in the 1990s' (Dalal-Clayton and Sadler 2005, 61). Procedural guidance for SEA was provided in *The Environmental Assessment Process for Policy and Programme Proposals* (FEARO 1993), with implementation subject to oversight by the Federal Environmental Assessment Review Office and later the Canadian Environmental Assessment Agency. In 1999, Canada reinforced its commitment to SEA with its release of the *1999 Cabinet Directive on the Environmental Assessment of Policy, Plan and Program Proposals*. It was not until January 2004, under an updated SEA Directive, that Canadian federal departments and agencies were required to prepare a public statement whenever a full SEA had been completed.

There are several guidelines available for SEA across Canadian federal departments and agencies. These guidelines reflect very different interpretations of SEA and what SEA is designed to accomplish. At the federal level, departmental and agency guidelines for SEA focus primarily on procedural support to ensure compliance with the SEA Directive. The directive outlines the guidelines and requirements for federal government departments on implementing SEA, which is operationalized at the department and agency level; it is assumed that as long as these guidelines are adhered to, the result will be meaningful SEA. The majority of the guidelines are designed for the appraisal of PPPs, such as those set forth by the Department of Foreign Affairs and International Trade or Transport Canada, although many of the guidelines identify classic EIA-based tools.

Outside the federal process, at the provincial level, SEA is practised largely on an ad hoc basis, and formal systems of SEA in Canada's provinces do not yet exist. SEA is interpreted as an extension of the regulatory environmental assessment process, such as under the Yukon Environmental and Socio-economic Board, as a separate review process for policy and program proposals, such as Newfoundland's *Strategic Environmental Review Guideline for Policy and Program Proposals*, or as part of project-specific terms of reference, such as in the province of Saskatchewan.

Aims and Objectives

SEA is based on the premise that project-based EIA, which in essence reacts to a proposed development, is not sufficient by itself to ensure the sustainable development of the environment. Consistent with the principles of Agenda 21, a series of actions aimed at greater sustainability and adopted at the 1992 Earth Summit by Canada and 177 other countries, SEA attempts to integrate *environment* into higher-order decision-making processes. Early integration of environment provides for a more proactive assessment process whereby alternatives are identified and assessed at an early stage

Table 13.1 Brief Timeline of SEA Development in Canada

1990	Policy reform on environmental assessment: Canadian Environmental Assessment Research Council releases guidelines for environmental assessment of policy and program proposals.
1991	Federal government reform package introduces Canada's first initiative in the development of a system of strategic environmental assessment: *Environmental Assessment in Policy and Program Planning: A Sourcebook.*
1992	North American Free Trade Agreement environmental review.
1993	FEARO procedural guidelines are released to federal departments on the EA process for policy and program proposals (FEARO 1993). Natural Resources Canada introduces guidelines for the integration of environmental considerations into energy policies.
1995	Amendments to the Auditor General Act require that all federal departments and agencies prepare a sustainable development strategy. Federal government releases *Strategic Environmental Assessment: A Guide for Policy and Program Officers.*
1996	Environmental assessment of the new minerals and metals policy.
1997	Department of Foreign Affairs tables Agenda 2000 outlining its commitment to environmental reviews of recommendations submitted to cabinet.
1999	Update to the 1990 cabinet directive is released and guidelines for implementation; SEA of Canada's commitment to Kyoto Protocol on greenhouse gas emissions is performed.
2000	Department of Foreign Affairs and International Trade prepares an SEA manual for federal program officers.
2001	Canada–Nova Scotia Offshore Petroleum Board publishes Canada's first SEA for the offshore oil industry. Transport Canada releases a policy statement demonstrating its commitment to and framework for the SEA of transport policies, plans, and programs. Department of Foreign Affairs and International Trade announces environmental assessment framework for trade negotiations.
2004	Guidelines on the cabinet directive on SEA are updated.
2007	Minister of Environment's Regulatory Advisory Committee, Subcommittee on SEA, commissions a report on the state of SEA models, principles, and practices in Canada.
2008	Canadian Council of Ministers of the Environment, Environmental Assessment Task Group, commissions a report to produce guidelines for regional SEA methodology and good practice.

Sources: Based on Noble 2002; 2003; 2008.

in the policy and planning process before irreversible decisions are taken (Box 13.3). Thus, SEA is simply a way of ensuring that downstream project planning and development occurs within the context of the *desirable* outcomes that society wants to achieve. In this way, SEA addresses the *sources* rather than the *symptoms* of environmental change.

Types of SEA

The specific nature of SEA will vary depending on the regulatory system and assessment context, but in general three types of SEA can be identified (Therivel 1993).

Box 13.3 SEA Benefits

- Streamlining project EIA by early identification of potential impacts and cumulative effects.
- Allowing more effective analysis of cumulative effects of broader spatial scales.
- Facilitating more effective consideration of ancillary or secondary effects and activities.
- Addressing the causes of impacts rather than simply treating the symptoms.
- Offering a more proactive and systematic approach to decision-making.
- Facilitating the examination of alternatives and the effects of alternatives early in the decision process before irreversible decisions are taken.
- Facilitating consideration of long-range and delayed impacts.
- Providing a suitable framework for assessing overall, sector-, and area-wide effects before decisions to carry out specific project developments are made.
- Providing focus for project EIA on how proposed actions fit within the broader context of the region.
- Verifying that the purpose, goals, and direction of a proposed plan or initiative are environmentally sound and consistent with broader policy, plan, and program objectives for the region.
- Saving time and resources by setting the context for subsequent regional and project-based EIAs, making them more focused, effective, and efficient.

Sources: Clark 1994; Cooper 2003; Kingsley 1997; Noble 2000; Sadler and Verheem 1996; Sadler 1998; Wood and Dejeddour 1992.

Policy-based SEA, often referred to as indirect SEA, is perhaps the most significant type of SEA, since large-scale government policies commonly have more far-reaching effects than individual development plans, programs, and projects. A number of policy SEA applications do exist, but their number is limited in comparison to sector-based SEAs, and many are policy appraisals rather than impact assessments per se. Canadian and international examples include the SEA of the North American Free Trade Agreement (see Hazell and Benevides 2000), Canadian minerals and metals policy SEA (see Noble 2003), and the SEA of two Danish bills under Denmark's EA system, one to amend laws relating to tenancy and housing conditions and rent subsidies and the second a subsidy scheme for private urban renewal (see Elling 1997). There are few publicly available policy-based SEAs in Canada. In 2007, an informal policy-based SEA was applied under the Biosphere Canada Action Program to assess national policy options for greenhouse gas mitigation in prairie agriculture (Box 13.4).

Sector-based SEA applies to sector-based initiatives, plans, and programs, such as forestry plans or oil and gas programs. The World Bank (1999) defines sector-based SEA as:

> an instrument that examines issues and impacts associated with a particular strategy, policy, plan, or program for a specific sector; including the evaluation and comparison of impacts against those of alternative options and recommendation of measures to strengthen environmental management in the sector.

Box 13.4 SEA of GHG Mitigation Policy Options in Prairie Agriculture

The Canadian agriculture and agri-food industries are significant producers of CO_2, NO_2, and CH_4 and responsible for approximately 7 per cent of total CO_2 equivalent emissions in Canada. Agricultural greenhouse gas (GHG) emissions, however, differ from those of other sectors in two respects. First, emissions from agriculture are for the most part not caused by energy production and use but rather through gas discharges from livestock production, soil disturbance, fertilizer use, and cropping practices. Second, the agriculture sector is also part of the solution to mitigating GHG emissions in that agriculture can provide a major carbon sink through enhanced soil management and improved cropping practices. Any single solution for GHG mitigation in the agriculture sector, however, may not necessarily be acceptable in the regulatory context or at the on-farm level. Moreover, a blanket mitigation policy may not be effective across all agricultural regions, creating gainers and losers because of variations in soil management practices, cropping systems, and resulting on-farm impacts.

In 2006, under the Biosphere Canada Action Program, a SEA was applied to evaluate competing policy options for GHG mitigation and to identify an appropriate strategy for prairie agriculture (see Noble and Christmas 2008). The assessment considered five alternatives, adopted from Agriculture and Agri-Food Canada's *Opportunities for Reduced Non-renewable Energy Use in Canadian Prairie Agricultural Production Systems* and from the *Canadian Economic and Emissions Model for Agriculture*, and was applied across the five major soil zones.

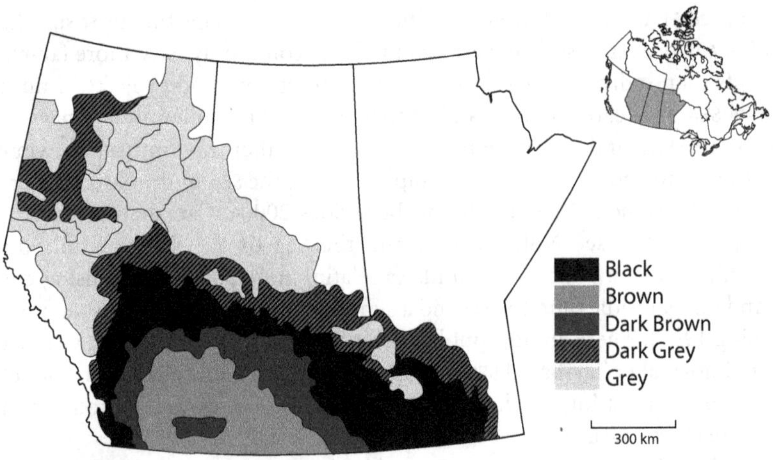

The policy alternatives are as follows:

A_1: Enhanced nitrogen-use efficiency
A_2: A 50 per cent increase in zero-tillage over current levels and direct seeding practices
A_3: A 50 per cent reduction in current summer fallow area
A_4: Shifting 10 per cent of the current cropland to forage production
A_5: A 10 per cent increase in the fuel efficiency of farm equipment

continued

A total of 13 VECs were identified against which the alternatives were assessed using a participatory, multi-criteria evaluation process:

Assessment VECs:

crop production quantity	crop production quality
economic risk	economic benefit
economic cost	flexibility of farm operations
institutional support	community support
time requirements	labour requirements
impact on soil resources	impact on water resources
complexity of mitigation	

The impacts of each mitigation policy option were assessed using five standard impact assessment characterization components: magnitude of the potential impact; direction of the expected impact; probability that the VEC would be affected by the proposed alternative; temporal duration of the potential impact; and management potential, or the capacity to offset the impacts of implementing the program given current levels of government or institutional support. In other words, the impact of GHG mitigation option 'A_n' on 'VEC_i' is determined as a function of the weight (w) of VEC_i and the magnitude (m), direction (d), probability (p), temporal duration (t), and management potential (mp), where the impact of option 'n' on VEC_i is defined as $w[d(m \times p \times t) \times mp]$. The total impact of any single policy option is thus defined as:

$$\sum_{i=1}^{n} w[d(m \times p \times t) \times mp]$$

Assessment data were standardized and evaluated using exploratory data analytical techniques to derive the *min-max* solution—the policy option that minimizes potential adverse impacts and maximizes the positive ones. The SEA results indicate considerable differences in the impacts of policy options, with mitigation option A_2 (increased use of zero-tillage cropping systems—an emerging practice in Canadian agriculture) identified as generating the most significant overall positive impacts. Using a concordance analysis, the aggregate results of the SEA indicate a strong preference for alternative A_2, zero-till practices, over all other competing GHG mitigation options: $A_2 >> A_4 > A_3 >> A_1 / A_5$, where '>>', '>', and '/' indicate 'strong outranking', 'outranking', and 'indifference', respectively.

The disaggregate assessment results, however, suggest considerable variation in impacts by soil type. While conservation and zero-tillage are gaining popularity across the prairie region, for example, it is primarily occurring in the moist, dark (black) soil zones. Summer fallow on the other hand—the practice of leaving a normally cultivated field free of vegetation for one growing season so as to conserve soil moisture—is identified by the Canadian Agricultural Census as a *declining* land management practice in the prairie region—a decline that is supported in the relatively moist dark brown and black chernozemic soil zones as a viable GHG mitigation option (A_3). However, increased zero-tillage and decreasing summer fallow are not the preferred mitigation options in other soil zones, in particular the brown chernozemic zone, where fallow is still relied upon as an important soil moisture conservation practice.

continued

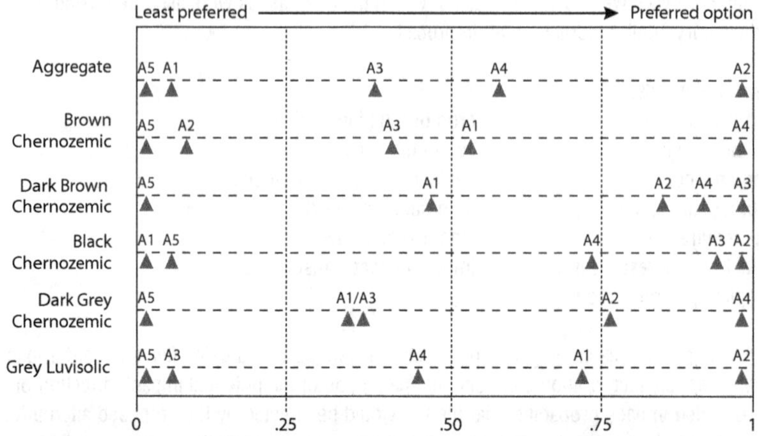

Results of the SEA indicate that a blanket policy for GHG mitigation may result in gainers and losers because of variations in soil type and soil management practices, cropping systems, and on-farm impacts. Regional programs, or satisficing solutions, established under a broader national policy and sensitive to soil characteristics and farm-level practices, may be more effective.

Source: Noble and Christmas 2008

Emphasis is placed on the initiatives of and alternatives to particular sector-based plans or programs that may lead to environmental change (Box 13.5). Initiatives and their alternatives are evaluated within the context of sector-based objectives, existing

Box 13.5 Benefits of Sector-Based SEA

- Avoids the limitations of project-specific EIAs in particular sectors by addressing broader environmental, social, and economic concerns.
- Prevents or avoids significant environmental effects through the assessment and development of sector-wide plans and programs before individual project decisions are taken.
- Provides opportunity for consideration of more effective or efficient sector plans or programs.
- Allows for planning and development of sector-wide environmental management strategies.
- Facilitates the inclusion of other sector interests in sector-wide planning and development.
- Provides a framework for considering the cumulative environmental effects of sector-wide development.

Source: World Bank 1999.

environmental conditions, and current and proposed plans and priorities. Sector-based SEAs are typically defined by jurisdictional or sector-based boundaries, such as the extent of sector activities or the area subject to a particular sector-based plan (e.g., forest harvest plan, urban master plan).

A number of sector-based SEAs have been undertaken in Canada in recent years, including forest management planning, electricity sector planning, and core area sector planning in Canada's national capital region (Box 13.6). Perhaps the sector in Canada where the most SEAs have occurred is Atlantic Canada's offshore oil and gas industry, under the Canada–Newfoundland and Canada–Nova Scotia offshore petroleum boards.

Box 13.6 Canada's National Capital Commission Core Area Sector Plan

In 2003, the National Capital Commission (NCC), an arm's-length federal Crown corporation with a mandate to plan lands in Canada's national capital region, commenced development of the Core Area Sector Plan (CASP). As part of plan development, the CASP was subject to SEA under the policy direction of the NCC and in the spirit of the Canadian cabinet directive. The SEA was conducted as a parallel process to plan development, with information feeding into the planning process with the intent to ensure that environmental considerations were built into future planning processes for the resulting strategies and projects. The SEA was situated within a tiered forward-planning system ranging from broad policy plans and overarching visions to site-specific objectives and implementation plans. As a sector plan, the CASP was third in a hierarchy of previous plans for the core area. Specific implementations and project-related actions under the CASP are to be addressed in subsequent project-based assessments under the Canadian Environmental Assessment Act. The purpose of the CASP was to identify a framework of policies and initiatives for development, programming, environmental integrity, transportation, animation, and architectural and design quality on federal lands in the capital's core area and to guide decision-making and inform planning initiatives over the next 20 years.

The SEA application was objectives-led and adopted a structured and systematic approach to assessing the implications of future initiatives resulting from the CASP according to specified biophysical, socio-economic, and cultural VECs and objectives. Potential future initiatives and projects under the plan were identified and each initiative reviewed to assess the potential environmental effects, including potential cumulative effects as a result of spatial and temporal crowding. Determination of significance was undertaken by cross-referencing the proposed CASP initiatives and other known activities in the region with foreseeable environmental conditions or trends. Mitigation measures and monitoring measures were recommended for each of the potentially adverse environmental effects. Alternatives to the plan were not explicitly assessed; rather, emphasis was placed on identifying a range of future planning actions or initiatives most likely to result from the CASP itself and evaluating their potential impacts. The NCC reports that the CASP has been improved through SEA application, and results of the SEA are reported to have improved subsequent planning initiatives. Public involvement and the systematic approach to assessment are identified as key strengths of the NCC's SEA methodology. However, since the SEA was conducted parallel to plan development, its results proved difficult to co-ordinate and integrate with the CASP planning process.

Source: Noble and Bronson 2007.

The most advanced system for SEA offshore, however, is under the UK Department of Trade and Industry (now the UK Department for Business, Enterprise and Regulatory Reform [BERR]), which has completed numerous SEAs for offshore licensing in recent years (Box 13.7).

Box 13.7 Department of Trade and Industry's Sector-Based SEA, Offshore UK

Oil and gas exploration and production in the UK is regulated through a petroleum licensing system. A production licence grants exclusive rights to the holder to search for, drill for, and extract petroleum in specified areas. The UK Department of Trade and Industry (DTI), prior to the UK SEA Directive coming into force, adopted a policy decision that SEA will be undertaken prior to future wide-scale licensing of the continental shelf and in 1999 began a series of sector-based SEAs to consider the implications of licensing for oil and gas activities. The first offshore SEA (SEA 1) was conducted in 1999–2000, with SEAs 2 through 7 following. Currently, an integrated offshore-energy SEA for further rounds of offshore licensing is being conducted. Details of the assessments completed to date can be found at http://www.offshore-sea.org.uk/site.

As an example, SEA 3 focused on parts of the central and southern North Sea, particularly 362 blocks, of which 30 are licensed, 205 had been under licences that are now relinquished, and 127 have not previously been licensed. The proposed initiative under consideration for the region was to offer additional production licences for blocks in the UK sector through the next round of offshore licensing. The alternatives to the proposed initiative were not to offer any blocks for additional licensing, to license only a restricted area, or to stagger the timing of licensing activity in the area.

The SEA considered a number of objectives and issues, including environmental protection objectives and standards established for the area through existing policies and plans, existing environmental problems that might be exacerbated by the proposed initiative, and likely impacts of the initiative and its alternatives, including potential incremental, cumulative, and synergistic impacts.

An initial scoping process with academic and conservation organizations commenced in early 2001. It was followed by a broader consultation exercise to identify main components and areas of concern related to the proposed initiative and assessment. A range of issues was identified, from socio-economic and environmental concerns to coastal defence and opportunities for co-ordinating offshore wind farm activity. The initiative and proposed alternatives were assessed on the basis of an initial environmental interaction matrix, expert judgment, stakeholder dialogue, and lessons from SEA 1 and SEA 2. Incremental effects were considered with regard to licensing actions likely to act additively with initiatives from other oil and gas activities, notably simultaneous and sequential surveying in existing and previously licensed areas. Cumulative effects were considered with regard to the potential impact of the initiative and alternatives on regional components in combination with the effects of other activities in the region of concern, notably seismic surveying and shipping, on fishing activities and marine resources.

Based on the SEA output and recommended mitigation measures, it was recommended that DTI proceed with the licensing initiative as proposed.

continued

Key Source/Effect	*Option 1: Not to offer any blocks*	*Option 2/3: Restrict the area temporally or spatially*	*Proposed Option: Proceed with licensing as proposed*
Noise, seabed damage, physical presence, discharges, emissions, waste to shore, accidents, cumulative effects, transboundary effects	No benefit or disadvantage	Potential but minor environmental effect or socio-economic disadvantage. No clearly identifiable or justifiable seasonal or spatial restrictions identified.	Potential but minor environmental effect or socio-economic disadvantage. No major effects predicted, given existing regulatory controls and mitigation.
Socio-economic effects	Potential but minor environmental effect or socio-economic disadvantage	Some benefit	Some benefit
Wider policy objectives	Potential significant environmental effect or socio-economic disadvantage	Strong benefit	Strong benefit

Source: Based on UK DTI 2002.

Often referred to as **regional-based** SEA, spatial SEA includes regional plans and programs, such as land-use planning, which may include multiple sectors. The World Bank (1999) defines regional-based SEA as:

> an instrument that examines the environmental issues and impacts associated with a strategy, policy, plan, or program for a particular region; including the evaluation and comparison of impacts against those of alternative options and recommendation of measures to strengthen environmental management in the region.

The purpose of regional-based SEA is to assess the impacts of plan and program initiatives within a particular region, in combination with other regional activities, in order to identify the preferred regional-based environmental planning or development strategy. The objective is to assist decision-making by systematically identifying a preferred option for regional management and development. For example, the Bow River Valley Corridor regional study in Alberta, while not explicitly labelled an SEA, placed considerable attention on evaluating the potential impacts of alternatives and competing land uses, including oil and gas projects, mining development, urban expansion, and agricultural grazing in the Bow River Valley. The outcome of the assessment assisted in the identification of preferred land-use development patterns and in setting limits to further development.

Regional SEA is often defined by environmental or ecological boundaries, may include multiple sectors, and is typically driven by environmental planning or management initiatives, state-of-environment reports, or the initiatives of and stresses caused by multiple sectors—an example is the regional environmental study of the Great Sand Hills, Saskatchewan (see Box 7.3). The Great Sand Hills assessment was based on an explicit SEA methodology consisting of a regional baseline assessment, trends analysis, and evaluation of alternative scenarios for future development in the region (see Noble 2008).

SEA FRAMEWORKS

Strategic environmental assessment has advanced considerably in recent years, including the development of new SEA frameworks. There is no single agreed-upon framework for SEA, but based on recent international experiences, a number of generic steps can be identified for SEA implementation. It is important to note that as with EIA, specific design requirements are often necessary within each application—including specific methods and techniques. Most of the methods and techniques required for SEA are readily available from project-level assessment and policy appraisal. The following sections present a brief outline of a generic SEA framework, including selected examples of methods and techniques. Much of the framework is modelled after EIA terminology and thus not explained in any great detail. The reader is encouraged to refer to previous chapters and to the glossary for a refresher on terms and concepts.

Develop a Reference Framework

Context-setting is essential to any environmental assessment application. Thus, the first phase in the SEA process is to establish the context (decision-making, policy,

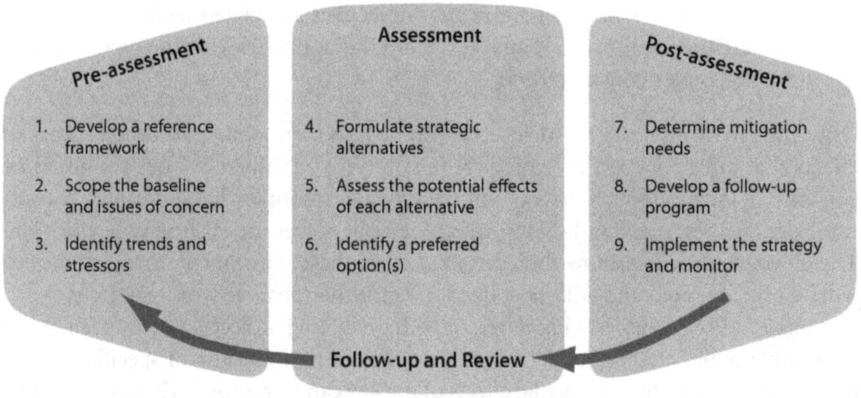

Figure 13.1 Strategic environmental assessment framework.

Source: Noble and Harriman 2009.

planning, regulatory) within which the assessment itself will take place. A number of basic yet fundamental questions should be asked at this stage, including:

- What are the objectives of the SEA, and what is hoped to be accomplished?
- What are the objectives of the strategic initiative or proposed plan?
- Who is responsible for the SEA, and what stakeholders are involved?
- What are the data requirements and information needs for carrying out the assessment?

In sector-based SEA, such as for offshore oil and gas, the assessment objectives may be more closely related to industry objectives for environmental performance, business initiatives, or industry and industry-related activities within the region, and thus the stakeholders and data requirements can be easily identified. For a broader regional-based SEA, such as the development of a regional land-use plan, the assessment objectives may focus on identifying an appropriate direction for longer-term sustainable regional development, including the implications of multiple-sector developments, such as housing, oil and gas exploration, and transportation, and interactions between these activities within the region.

Scope the Baseline and Issues of Concern

Once the purpose of the SEA is determined, the current baseline must be established and primary issues of concern relevant to the sector, region, or policy environment identified. Depending on the tier of assessment (policy, plan, or program), emphasis may focus on broader policy issues and the policy environment or on the state of the region and on particular environmental VECs. Generally speaking, establishing the baseline involves asking:

- What are the VECs of concern?
- What is the current (and cumulative) condition of potentially affected VECs?

- What are the thresholds of concern or objectives associated with the VECs?
- What are the appropriate spatial and temporal boundaries for the type of region or sector under consideration?

Only the environmental or socio-economic components likely to be receptors of effects that are important to ecological functioning or valued by society should be considered, including such VECs as air quality, water resources, habitats, spaces of cultural significance, social well-being, or, at the policy level, an existing policy or plan. The objective is to determine the 'current state-of-the-environment' in which the PPP will be implemented and will most likely affect, either positively or negatively.

Establish VEC indicators or objectives. As with predicting project-based impacts, making useful predictions about effects in SEA requires some 'indicator' or specified 'objective' against which the incremental effects of the PPP can be assessed. Thus, an indicator should define each VEC that is scientific in nature and capable of providing an early warning of the state or health of the VEC of concern. In cases where VECs are not amenable to indicator selection, or when dealing with broader policy-based issues, it is more appropriate to focus on VEC objectives or criteria against which potential changes in VEC conditions, or anticipated outcomes of the PPP, can be appraised. These objectives or criteria can be derived or translated from previously stated goals and objectives or set within the context of a broader environmental vision for the region or sector of concern. Guidelines for VEC indicator selection are discussed in Chapter 6, as are the various approaches to establishing VEC criteria or objectives, namely:

- absolute ecological or socio-economic thresholds;
- acceptable limits of change;
- desired conditions or outcomes.

Delineate assessment boundaries. Spatial and temporal boundaries provide a frame of reference for SEA and help to establish the resolution of analysis. Spatial and temporal bounding for SEA is no different, in principle, from that for project-based EIA. The primary difference at the strategic level concerns the nature of the activities that must be considered. For example, geographic relationships, common resources, and proposed activities must be viewed not only from the perspective of physical and socio-economic phenomena but also from the perspective of current and proposed policies, plans, or programs that may interact with the proposed plan or program.

Identify Trends and Stressors

This is the retrospective phase of SEA and serves to identify the key trends and driving forces of change in the sector, region, or policy environment of concern. It may require identifying cause–effect relationships or correlations between VEC conditions and various actors or drivers of change. In some cases, particularly at the broader policy level, specific causal relationships may be difficult if not impossible to define. In such cases, attention should focus on identifying past policy initiatives and policy changes and recognized environmental or public responses. The objective is not to spend a great deal of time and resources on identifying ecosystem-wide linkages, for example, but rather to gain an understanding of a limited number of important components and possible interactions such that key trends or observations from the baseline can be carried forward to the future and used as a basis against which alternative options are assessed and appraised. A variety of methods and techniques are available to support trends and stressors analysis, including retrospective modelling, network analysis, and photographic progressions.

Develop Strategic Alternatives

Unless there is more than one potential and feasible way to proceed, no decision has to be made, and therefore no SEA is required; this includes the option of no plan or action. Alternatives in SEA should include:

- the current baseline condition carried forward;
- the proposed initiative, plan, or plan or program implementation procedure;
- alternatives to the initiative, proposed plan, or plan or program implementation procedure.

This process might include the construction of scenarios for each alternative that emphasize different:

- spatial and temporal attributes or option implementation;
- modes or processes, including technologies or methods;
- means of meeting the plan or strategic initiatives;
- targets or objectives.

Alternatives should be considered only if they are:

- consistent with broader environmental and sustainability goals and objectives;
- not in conflict with existing regional or sector policies, plans, or activities; and/or
- economically, technologically, and institutionally feasible.

The objective is to identify alternatives that are 'more sustainable' or 'least negative' or that trigger the 'least significant' amount of environmental change. If, in the case of a proposed plan or program, a strategic alternative is likely to cause a 'more significant adverse effect' than the proposed plan or program itself, then it should not be considered a viable alternative. Thus, for each alternative, the following minimum factors should be considered in order:

- Is the alternative compatible with sustainability goals and objectives?
- Is there a potential interaction (either positive or negative) between activities associated with the proposed alternative and existing VECs?
- Are potentially affected VECs already affected (either positively or negatively) by existing policies, plans, or activities in the sector or region?
- Will the alternative have an effect on current VECs (either positive or negative) in combination with already existing and proposed policies, plans, and activities?

Table 13.2 gives an example of how options might be scoped in light of potential interactions with existing regional or sector plans and activities.

Table 13.2 Simple Checklist of a Preliminary Alternatives Scoping Matrix

Plan options or alternatives	*Is the alternative compatible with existing and proposed policies, plans, and activities?*	*Is there a potential for interaction with and effect on current VECs?*		*Is the VEC already affected by existing plans or activities?*	*Is there a potential for cumulative interaction?*	*Should the alternative be given further consideration?*
A1		water quantity				☐ Yes
		hunted species				☐ No
		potable water				☐ Uncertain
A2		water quantity				☐ Yes
		hunted species				☐ No
		potable water				☐ Uncertain
A3		water quantity				☐ Yes
		hunted species				☐ No
		potable water				☐ Uncertain

Source: Based on the UK Office of the Deputy Prime Minister 2003.

Assess the Potential Effects of Each Alternative

This is the prospective or futures-oriented phase of SEA. Only those options or alternatives that are compatible with the goals and objectives of existing and proposed policies, plans, and initiatives for the sector or region should be considered for detailed assessment. That way, the focus of assessment is on the *desirability* of the initiatives with regard to VEC objectives rather than on trying to resolve conflicting goals and objectives between proposed and existing initiatives.

The assessment of options should include consideration of whether:

- the proposed option will have an effect on specified VECs;
- there will be potential cumulative effects on VECs;
- other policies, plans, or actions may affect the same resources;
- the effects are likely to be significant.

Determine environmental changes likely to affect VECs. The analysis of potential effects involves the prediction of the potential effects of each option on the VECs by identifying the primary sources of stress and associated effects pathways. This is a formidable task, and the levels of complexity and uncertainty increase as one moves from projects to plans and from sectors to regions. The objective is to identify potential stressors and VEC responses associated with each alternative. The level of detail depends on the availability of data.

Identify VEC response. This requires understanding the present conditions of VECs and predicting how each might react based on the specified VEC indicators. In short, the objective is to capture, based on the relational networks established, how the VEC receptor might respond or deviate from its current condition and future baseline condition given the multiple interactions and pathways with and without each of the proposed options. Depending on the nature of the VEC receptor and level of information available, it may be possible to model trends over time based on different scenarios of stressors and the range of VEC responses for each option; in other cases, expert-based forecasting, such as the Delphi approach, might be most appropriate. In general, such futures methods can be readily adapted from EIA-driven and planning approaches (Box 13.8).

As baseline data and knowledge of cause–effect relationships increase, it is possible to adopt more 'science-driven' methods and techniques for effects assessment; in cases where data are lacking or where an understanding of cause–effect relationships is simply not possible, more 'judgment-driven' approaches must be replied on. Duinker and Greig (2007) and Cherp, Watt, and Vinichenko (2007) also emphasize the need to consider 'external wildcards' (e.g., climate change) and 'emergent and external' events (e.g., policy influences, economic change) that may affect VEC responses or the desirability of any given alternative or future scenario.

Identify a Preferred Option

The outcome of a SEA should lead to the identification of a desirable strategic direction. In other words, when compared to all other proposed options, a preferred strategy for action is identified on the basis of its:

Box 13.8 Selected Methods and Techniques Suitable to SEA of Alternatives

Impact models: flow diagrams; loop diagrams; network diagrams; quantitative modelling

Geographic Information Systems: spatial relationships; scenarios over time and space

Futures methods: trend analysis; extrapolation; scenario development

Expert judgment and valuation: forecasting; Delphi approaches; weighting and scoring

Econometrics: Monte Carlo simulation; national income accounting

Risk assessment: probability of outcomes; risk pathways; risk perception

- potential effects on specified VECs;
- potential for cumulative effects on VECs;
- interactions with and the contributions of current and reasonably foreseeable policies, plans, and initiatives within the sector or region, with or without the proposed plan or initiative.

Evaluate alternatives. For SEA, the issue may not always be about 'predicting' and impacts as much as it is about identifying preferred directions and outcomes. In other words, once environmental effects are identified, they must be evaluated. The question now becomes 'what is required to achieve a desired future or VEC outcome, and what are the consequences of different choices?' VEC or plan goals and objectives, such as minimizing the potential for effects and cumulative interactions between plans or programs affecting water quality, for example, and the alternative means of achieving these goals and objectives are evaluated. Thus, emphasis might be placed on assessing the feasibility and desirability of meeting a desired target (e.g., maintaining a specified level of biodiversity) or on selecting alternatives that minimize potential cumulative effects or maximize environmental and socio-economic contributions.

For each VEC objective, one should consider how *desirable* strategic option '*i*' is compared to option '*j*' when taken into consideration with the potential effects of current and reasonably foreseeable policies, plans, and initiatives. In the case of SEA for GHG mitigation policy for prairie agriculture (Box 13.4), for example, alternatives were evaluated on the basis of whether they met a range of VEC objectives simultaneously, including maximizing on-farm flexibility in policy implementation and minimizing economic risk. In the case of the Great Sand Hills assessment in Saskatchewan (Box 7.3), the effects of alternative scenarios of development were ultimately evaluated according to their ability to meet specified targets and objectives for biodiversity conservation.

How the options are evaluated in light of the objectives again depends on the availability of data and desired decision resolution and may include, for example, expert-based judgment or more complex *modelling* and *simulation analysis*. In keeping with the examples and decision-making techniques discussed at various points throughout this text, Box 13.9 presents an example of alternatives evaluation and decision-making based on multi-criteria evaluation and expert judgment.

Box 13.9 Example of Alternatives Evaluation in SEA Based on the Paired Comparison Weighting Technique

(Refer to Chapter 8, Box 8.1, and Chapter 12, Box 12.3, for an explanation of the paired comparison technique.)

How desirable is strategic option 'i' compared to option 'j' when taken into consideration with the effects of current and reasonably foreseeable policies, plans, and initiatives on the specified VEC?

VEC objective 'n'	*Option 1*	*Option 2*	*Option 3*
Option 1	1	1/3	7
Option 2	3	1	7
Option 3	1/7	1/7	1

Where:

9 = option 'i' is extremely more desirable in comparison to option 'j'
7 = option 'i' is strongly more desirable in comparison to option 'j'
5 = option 'i' is moderately more desirable in comparison to option 'j'
3 = option 'i' is slightly more desirable in comparison to option 'j'
1 = option 'i' is equally desirable to option 'j'
1/3 = option 'i' is slightly less desirable in comparison to option 'j'
1/5 = option 'i' is moderately less desirable in comparison to option 'j'
1/7 = option 'i' is strongly less desirable in comparison to option 'j'
1/9 = option 'i' is extremely less desirable in comparison to option 'j'

↑ Compared to option 'j', option 'i' is more likely to: a) meet the VEC objective and/or b) minimize overall effects on the VEC.

↓ Compared to option 'j', option 'i' is less likely to: a) meet the VEC objective and/or b) minimize overall effects on the VEC.

Normalized matrix:

VEC objective 'n'	*Option 1*	*Option 2*	*Option 3*
Option 1	0.24	0.23	0.47
Option 2	0.72	0.68	0.47
Option 3	0.03	0.10	0.07

Priority vector : Option 1 = 0.31; Option 2 = 0.62; Option 3 = 0.07

$$0.31 \begin{bmatrix} 1 \\ 3 \\ 1/7 \end{bmatrix} + 0.62 \begin{bmatrix} 1/3 \\ 1 \\ 1/7 \end{bmatrix} + 0.07 \begin{bmatrix} 7 \\ 7 \\ 1 \end{bmatrix} = \begin{bmatrix} 1.01 \\ 2.04 \\ 0.21 \end{bmatrix}$$

Rank alternatives based on VEC objective 'n' (divide the vector by original priorities):

Option 1 = 1.01/ 0.31 = (3.26)
Option 2 = 2.04/ 0.62 = (3.29)
Option 3 = 0.21/ 0.07 = (3.00)

Option 2 (3.29) > Option 1 (3.26) > Option 3 (3.00)

continued

Normalize the ranking by $[(i - i_{min}) / (i_{max} - i_{min})]$ to display the relative 'magnitude' of the ranking for each VEC objective:

(Option 1 = 0; Option 2 = 1; Option 3 = 0.10)

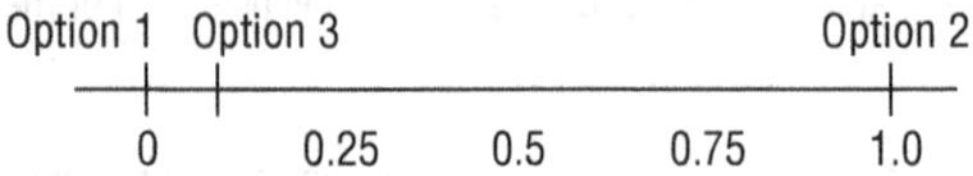

Determine 'cumulative desirability' of the option:
Repeat the above procedures for each VEC objective. The 'cumulative desirability' and final preferred option or initiative is determined by summing the normalized rankings across all VEC objectives. For example:

Objectives	*Option 1*	*Option 2*	*Option 3*
VEC objective 1	0.00	1.00	0.10
VEC objective 2	0.00	1.00	0.24
VEC objective 3	0.36	0.00	1.00
VEC objective 4	0.11	0.00	1.00
'Cumulative desirability'	0.47	2.00	2.34

The net result is a ranking of all options and identification of two competing options (Option 3 and Option 2), of which Option 3 is ultimately the preferred choice based on the (unweighted) objectives across all VECs. To determine the weighted rankings, simply multiply the option scores by VEC weights (importance) prior to normalizing. In this particular example, given the proximity of Option 3 and Option 2, a sensitivity analysis to changing VEC objectives and priorities would be warranted.

Determine Mitigation Needs

Even the 'preferred' option may have potentially adverse environmental effects that cannot be avoided. A number of questions can be examined across VECs to determine whether effects associated with strategic initiatives require mitigation, notably:

- Will the proposed initiative or plan option generate effects that will exceed thresholds as defined in regulations or guidance or as established by maximum allowable effects levels?
- Are the effects significant when considered in light of the current conditions of the affected VEC?
- Will the plan or plan option generate effects on VECs that are likely to interact cumulatively with other plans or activities in the region on already stressed VECs?
- Will the effects of the plan or plan option be permanent?
- Does the initiative or plan option conflict with the goals and objectives of existing plans and activities?

Attention should also be given to managing current conditions such that the preferred option can be implemented. For example, current industry or land-use conflicts may need to be resolved before the preferred option can become viable. Once these questions are addressed for the 'preferred' option, its preference should be re-evaluated to determine whether it is still the preferred option based on the mitigation requirements and cost and feasibility of implementing them. Competing options may need to be considered and a satisfying option pursued.

Develop a Follow-up and Monitoring Program

No assessment process is complete without a follow-up or monitoring component. The post-decision activities of follow-up and monitoring are essential to ensuring that SEA, and the resulting PPP, is delivering its anticipated or desired outcomes and that any mitigation measures prescribed have been implemented and are working. The rationales for follow-up at the strategic tier are generally the same as of those at the project level: derived from the notions of uncertainty and risk associated with decision-making and on the need for feedback and learning. However, the complexities and uncertainties of determining post-implementation environmental implications are exacerbated at the strategic level in that:

- PPPs are often formulated in abstract terms, resulting in vague directions for acting;
- whereas significant deviations from the original plan are abnormal at the project level, they are typical of strategic-level processes;
- there are fewer direct linkages between decisions at the strategic level and actual PPP impacts (Cherp, Watt, and Vinichenko 2007).

Depending on the nature and purpose of the SEA—to appraise a PPP or to assess the impacts of a PPP—follow-up should be designed, at a minimum, to track PPP performance on the basis of established goals, objectives, and targets or to track environmental impacts and the effectiveness of management measures, as well as to improve scientific and technical knowledge through feedback.

There are three broad types of follow-up and evaluation in SEA:

1. **Input evaluation** concerns SEA procedures and requirements, such as SEA purpose and objectives, data quality, linkages to PPP initiatives, and ensuring that these requirements and objectives were met.
2. **Process evaluation** concerns the SEA application, such as impact predictions and mitigation measures, and following up to verify the accuracy of these impacts and to determine whether the mitigation measures were implemented and effective.
3. **Output evaluation** concerns SEA effectiveness and involves determination of whether the analysis informed the PPP decision process, whether the proposed PPP was modified as a result of the SEA, whether the SEA influenced downstream planning and project impact assessment, and whether what were intended as plan or initiative outcomes were in fact realized.

Implement the Strategy and Monitor

Regardless of the process efficiency and effectiveness, the output of an SEA is of little benefit if it is not implemented. However, implementation of a new strategy or

PPP is easier said than done. The implementation of a strategic initiative requires a degree of collaboration that extends beyond the ability of any single government agency or department. Successful implementation demands a level of commitment and interagency collaboration that is not common in impact assessment.

There is no single best PPP implementation style; rather, implementation depends on the environment within which the PPP is being implemented (Table 13.3) (Noble and Harriman 2008). Generally speaking, however, the more complex the strategy and the greater the uncertainty or potential conflict involved, the more preferred an adaptive approach to implementation is. It is important to be realistic about the limitations and uncertainties in looking into the future and in developing blueprint implementation and management strategies. Strategies and PPPs must be sufficiently adaptive to system changes, bifurcation, and external and emergent stressors and responsive to new knowledge gained through monitoring and follow-up processes (Cherp, Watt, and Vinichenko 2007).

DIRECTIONS IN SEA

Strategic environmental assessment is still quite new and relatively limited in terms of its adoption and application. A large number SEA applications do not occur under the SEA nametag or under any form of environmental assessment framework. However, based on known experiences to date, Noble (2009) describes Canadian SEA systems and practices as diverse, far from consolidated in scope and function, and encompassing a range of models and practices. As such, there is considerable variability in outcome and expectations.

There is no single model of SEA that can be unequivocally applied; rather, attention must be given to context, to regulatory opportunities and constraints, and to the tier of application. This is not to say that 'good-practice' SEA should not be defined but that SEA operates in diverse forms and under a range of institutional arrangements. Common knowledge and understanding of the SEA process and what SEA is supposed to deliver, the influence of SEA over subsequent assessment and decisions,

Table 13.3 Adaptive versus Programmed Approaches to Alternative or PPP Implementation

	Strategy or PPP environmental context	
Guidelines	*Structured*	*Unstructured*
scope of change	incremental	major
certainty in strategy, technology, or management	certain within risk	uncertain
conflict over goals and means	low conflict	high conflict
structure of institutional setting	tightly coupled	loosely coupled
stability of environment	stable	unstable
Preferred implementation style	*Programmed*	*Adaptive*

Source: Based on Berman 1980, from Mitchell 2002.

and an institutional commitment and willingness to carry out SEA are among the most significant challenges to its continued development in Canada. To date, the Canadian SEA report card might read as follows: considerable promise, falling short of its potential (Noble 2009).

KEY TERMS

input evaluation
output evaluation
plan
policy
policy-based SEA
process evaluation
program
programmatic environmental assessment
regional-based SEA
sector-based SEA
strategic environmental assessment

STUDY QUESTIONS AND EXERCISES

1. What are the potential benefits and challenges of applying environmental assessment to policy?
2. Identify the different types of SEA, and provide an example of the type of problem or situation to which each might apply.
3. What are the provisions, if any, for policy, plan, or program assessment in the political jurisdiction in which you live? What do you see as the main challenges to SEA where you live?
4. Visit the websites for the Canadian cabinet directive for implementing SEA at www.ceaa.gc.ca and the European directive at http://europa.eu.int/comm/environment/EIA/home.htm. Compare and contrast the directives in terms of their objectives, scope, requirements for reporting, and public involvement.
5. An important part of SEA is the identification of alternatives. One question that emerges from this, however, is who should be involved. Consider two hypothetical plan proposals in your area, one for oil and gas licensing and one for the development of a regional land-use plan. Discuss who should be involved in the identification of alternatives.
6. Obtain a completed SEA from your local library or government registry, or access one online. Using Box 13.1 as a guide, explore the SEA for evidence of 'strategic' characteristics. In other words, are the characteristics listed in Box 13.1 present in the SEA? Are there certain characteristics that seem to be missing? Compare your findings to those of others.

REFERENCES

Berman, P. 1980. 'Thinking about programmed and adaptive implementation: Matching strategies to situations'. In H. Ingram and D. Mann, Eds., *Why Policies Succeed or Fail*. Beverley Hills, CA: Sage.

Cherp, A., A Watt, and V. Vinichenko. 2007. 'SEA and strategy formation theories: From three Ps to five Ps'. *Environmental Impact Assessment Review* 27: 624–44.

Clark, R. 1994. 'Cumulative effects assessment: A tool for sustainable development'. *Environmental Impact Assessment Review* 12 (3): 319–22.

Cooper, L. 2003. *Draft Guidance on Cumulative Effects Assessment of Plans*. EPMG Occasional Paper 03/LMC/CEA. London: Imperial College.

Dalal-Clayton, B., and B. Sadler. 2005. *Strategic Environmental Assessment: A Sourcebook and Reference Guide to International Experience*. London: Earthscan.

Duinker, P., and L. Greig. 2007. 'Scenario analysis in environmental impact assessment: Improving explorations of the future'. *Environmental Impact Assessment Review* 27: 206–19.

Elling, B. 1997. 'Strategic environmental assessment of national policies: The Danish experience of a full concept assessment'. *Project Appraisal* 12 (3): 161–72.

FEARO (Federal Environmental Assessment Review Office). 1993. *The Environmental Assessment Process for Policy and Program Proposals*. Hull, QC: Minister of Supply and Services Canada.

Hazell, S., and H. Benevides. 2000. 'Toward a legal framework for SEA in Canada'. In M.R. Partidário and R. Clark, Eds., *Perspectives in Strategic Environmental Assessment*. New York: Lewis Publishers.

Kingsley, L. 1997. *A Guide to Environmental Assessments: Assessing Cumulative Effects*. Hull, QC: Parks Canada, Heritage Canada.

Mitchell, B. 2002. *Resource and Environmental Management*. 2nd edn. Harlow, UK: Prentice Hall.

Noble, B.F. 2000. 'Strategic environmental assessment: What is it and what makes it strategic?' *Journal of Environmental Assessment Policy and Management* 2 (2): 203–24.

———. 2002. 'Strategic environmental assessment of Canadian energy policy'. *Impact Assessment and Project Appraisal* 20 (3): 177–88.

———. 2003. 'Auditing strategic environmental assessment in Canada'. *Journal of Environmental Assessment Policy and Management* 5 (2): 127–47.

———. 2008. 'Strategic approaches to regional cumulative effects assessment: A case study of the Great Sand Hills, Canada'. *Impact Assessment and Project Appraisal* 26 (2): 78–90.

———. 2009. 'Promise and dismay: The state of strategic environmental assessment systems and practices in Canada'. *Environmental Impact Assessment Review* 29 (1): 66–75.

Noble, B.F., and J. Bronson. 2007. *Models of Strategic Environmental Assessment in Canada*. Ottawa: Canadian Environmental Assessment Agency.

Noble, B.F., and L. Christmas. 2008. 'Strategic environmental assessment of greenhouse gas mitigation options in the Canadian agricultural sector'. *Environmental Management* 41: 64–78.

Noble, B.F., and J. Harriman. 2008. *Regional Strategic Environmental Assessment: Methodological Guidance and Good Practice*. Report prepared for the Canadian Council of Ministers of the Environment. Winnipeg.

———. 2009. 'Strategic environmental assessment'. In K. Hanna, Ed., *Environmental Impact Assessment: Practice and Participation*. Don Mills, ON: Oxford University Press.

Partidário, M. 2007. *Strategic Environmental Assessment Good Practices Guide: Methodological Guidance*. Lisbon: Portuguese Environment Agency.

Sadler, B. 1996. *International Study of the Effectiveness of Environmental Assessment, Final Report, Environmental Assessment in a Changing World: Evaluating Practice to Improve Performance*. Canadian Environmental Assessment Agency and International Association for Impact Assessment. Ottawa: Minister of Supply and Services Canada.

———. 1998. 'Ex-post evaluation of the effectiveness of environmental assessment'. In Alan L. Porter and John J. Fittipaldi, Eds., *Environmental Methods Review: Retooling Impact Assessment for the New Century*. Fargo, ND: The Press Club.

Sadler, B., and R. Verheem. 1996. *Strategic Environmental Assessment: Status, Challenges and Future Directions*, Report 53. Ottawa: Canadian Environmental Assessment Agency.

Therivel, R. 1993. 'Systems of strategic environmental assessment'. *Environmental Impact Assessment Review* 13: 145–68.

UK DTI. 2002. *Strategic Environmental Assessment of Parts of the Central and Southern North Sea: SEA 3.* UK Department of Trade and Industry.

UNEP (United Nations Environment Programme), Economics and Trade Program. 2002. *Environmental Impact Assessment Training Manual*. 2nd edn. New York: UNEP.

Wood, C., and M. Djeddour. 1989. 'Environmental assessment of policies, plans and programs'. Interim report to the Commission of European Communities. Manchester, UK: EIA Centre.

_____. 1992. 'Strategic environmental assessment: EA of policies, plans and programmes'. *Impact Assessment Bulletin* 10 (1): 3–23.

World Bank. 1999. *Environmental Assessment*. Operational Policy and Bank Procedures, no. 4.01. Washington: World Bank.

World Commission on Environment and Development. 1987. *Our Common Future*. Oxford: Oxford University Press.

PART V

Environmental Impact Assessment Postscript

CHAPTER 14

The Effectiveness of Environmental Assessment: Retrospect and Prospect

INTRODUCTION

Conceived initially as a pragmatic tool for ensuring that at least some forethought and foresight are given to the impacts of a proposed development project before that project becomes a reality, EIA is now an accepted part of development decision-making—but has it been a worthwhile part? Would decisions about development have been different and the environmental effects more severe if no EIA had taken place? Has EIA really made a difference? These are challenging questions for the environmental assessment community, and the answers demand a comparison of development outcomes with EIA to those without EIA. In that sense, the answers may perhaps be more hypothetical and argumentative than factual. Over the years, however, there have been several inquiries into the effectiveness of EIA and whether EIA is making a difference. The majority of these inquiries have focused on the procedural aspects of EIA, from screening and scoping to significance determination and post-decision environmental monitoring. The view is that by improving the process of EIA, its benefits will be more fully realized (Jay et al. 2007). This book, emphasizing EIA procedure, admittedly reflects that view: there does remain considerable room for EIA process improvement. Perhaps the real measure of EIA, however, is its substantive effectiveness—its contribution to environmental management and longer-term sustainability. This chapter reflects briefly on the substantive effectiveness of environmental assessment in Canada, as illustrated by the current state of project-based EIA, CEA, and SEA systems and outcomes, and introduces the case for a more regionalized and integrated approach to impact assessment.

THE SUBSTANTIVE EFFECTIVENESS OF ENVIRONMENTAL ASSESSMENT

Environmental assessment has evolved considerably since its inception. As Gibson (2002) explains, the concept and practice has moved towards being conducted earlier in the planning process, more open and participative, more comprehensive, more mandatory, and more closely monitored. Environmental assessment has also evolved in a more substantive way, including efforts to 'link up' with environmental management, to assess cumulative environmental effects, and to advance application to the strategic tier. At the same time, however, there appears to be a growing dissatisfaction with environmental assessment and increasing pressures on the environmental assessment practitioner and academic community to respond (see Boyden 2007). Surveying the environmental assessment literature, it would appear that project-based EIA, CEA, and SEA are falling far short of expectations.

Environmental Impact Assessment

Environmental assessment first appeared on the scene in the early 1970s. Now, thousands of EIAs are completed on an annual basis. The underlying concept of EIA is quite simple—identifying, assessing, and finding ways of mitigating the potential impacts of proposed projects on the human and biophysical environment. The subject of assessment is typically an individual infrastructure project and its potential effects. Procedurally, when a project is proposed and the need for an EIA determined, the most likely impacts of that project are assessed. A decision is then made as to whether the project might cause significant adverse environmental effects and whether such effects can reasonably be mitigated. The philosophy of EIA at the time of its establishment, and still very much so today, is one of a rationalist model requiring a technical evaluation of project design and of the local environment in order to provide *sound* advice to proponents and decision-makers. In its early years, much of this advice was about pollution control through engineering and abatement technologies. In recent years, however, this sphere has expanded to include socio-economic and cultural issues, benefits-sharing, sustainability, the assessment of cumulative environmental effects, and an expectation that EIA also contribute to ongoing environmental management (e.g., Storey 1986; Bailey 1994).

That said, and as highlighted in Chapter 1, Richard Fuggle, president of the International Association for Impact Assessment (IAIA) at the time, described in an IAIA newsletter a 'disillusionment' with EIA '. . . and scepticism that impact assessments are contributing to better decisions' (Fuggle 2005, 1). Noble and Storey (2005) reported on the utility of follow-up in EIA and on the limited attention to social impact management and monitoring post-development decisions (see Chapter 10). More recently, Galbraith, Bradshaw, and Rutherford (2007) discussed the rise of privatized environmental and socio-economic impact benefit agreements between communities and private industry—perhaps a response, at least in part, to the failings of the EIA process to adequately consider benefits or build trust and capacity among stakeholders (see Chapter 9). Clearly, whether EIA is delivering on its expectations is at issue.

Cumulative Effects Assessment

A significant development in the evolution of EIA was recognition of the cumulative nature of environmental effects and the need for some form of CEA. The need to better assess and manage cumulative environmental effects really took shape in Canada during the early 1980s, and by the 1990s it had become a requirement under the Canadian Environmental Assessment Act for all project impact assessments (see Chapter 12). In practice, however, the effects of development continued to be assessed and managed on a project-by-project basis, with little regard for the effects that might result in combination with other past, present, and foreseeable development activities. Recent reviews of the state of CEA in Canada, for example, suggest that CEA is simply not working; it remains narrow, reactive, and divorced from the broader planning and decision-making context (e.g., Harriman and Noble 2008; Dubé 2003). Impacts are assessed on a stressor-by-stressor basis, and impact statements typically refer to cumulative effects in a separate section, the assumption being that cumulative effects are a different class of effects, derived from summing the total effect of individual stressors. Duinker and Greig (2006) go as far as suggesting that CEA in its current form is doing more harm than good. Part of the challenge is that CEA simply cannot work within the confines of project-based EIA, and properly assessing and managing cumulative environmental effects is well beyond the scope and scale of any individual project proponent (see Ross 1994; Creasey 2002). Under project impact assessment, cumulative effects are considered only in relation to the incremental effects of a proposed activity rather than as the cumulative effects of everything on the *valued ecosystem components* of concern.

One response to the constraints of doing CEA inside project EIA has been the rise of regional studies in which CEA adopts a more *effects-based* approach and focuses on understanding environmental systems and relationships rather than focusing on individual project stressors. Several regional CEA studies have unfolded in Canada over the past 15 years, and regional frameworks for CEA have been the focus of several Canadian Environmental Assessment Agency research and development programs. The contribution of regional CEA studies to solving the problems of project-based CEA, however, has not materialized. As Harriman and Noble (2008) explain, regional frameworks for CEA have value in their own right, but they have operated at a different spatial scale from that of project EIA, tend to address different VECs and indicators, and occur outside the environmental assessment system as a parallel study with no real requirement for their integration or adoption in development decision-making. According to Spaling et al. (2000), rarely is there authority to implement recommendations or to carry forward regional effects–based findings to EIA, and because regional studies are conducted in many different jurisdictions and for different purposes, it is unlikely that a consistent model will be developed or utilized (Grzybowski and Associates 2001).

Strategic Environmental Assessment

SEA was introduced in Canada in the early 1990s and became more formally established in 1999 under a federal cabinet directive. SEA was touted as the solution to addressing area-wide and cumulative effects problems, and it was assumed that applying SEA early in the decision-making process would allow sustainability benefits to trickle down to the project level. No formal systems of SEA exist in Canada outside the current federal cabinet directive; however, various forms of SEA have been ongoing in Canada for a number of years, both formally and informally and under a variety of labels and institutional models. That said, a recent review of the state of SEA systems and models in Canada concludes that SEA is diverse, founded on a range of principles and frameworks, and not well understood (Noble 2009). The result is considerable variability in SEA experience and added value, largely because of the institutional and methodological pluralism of SEA. Noble characterizes the current state of SEA as one of 'promise and dismay', going on to note that 'under the federal system . . . many applications have been disappointing . . . some of the better examples of SEA have neither carried the SEA nametag nor occurred under formal SEA requirements . . . of particular concern is the systematic separation of SEA from downstream decision inputs and assessment activities.' The benefits of SEA in facilitating CEA have not been forthcoming.

INTEGRATING THE SILOS OF ENVIRONMENTAL ASSESSMENT

Does this mean that environmental assessment in Canada has not been delivering—that it has not been effective? On the surface, the answer may appear to be an overwhelming yes; however, the question is by no means superficial. Arguably, there are simply too many different expectations of each model of impact assessment—or perhaps the right things are expected but from the wrong process. For example, one cannot expect regional understandings of cumulative effects to emerge from within the confines of project-based EIA—yet project proponents are required by law at the federal level to assess cumulative environmental effects beyond the scope and scale of their project and then are heavily criticized for their approach and the quality of the results. At the same time, SEA cannot be expected to replace project-based impact assessment or to deliver the data required by project proponents to get their projects approved, even though project proponents are increasingly calling for a more streamlined environmental assessment and approvals process. Nor can SEA deliver on CEA in the absence of the regional models and institutional frameworks necessary to implement and sustain it.

Part of the challenge to the effectiveness of environmental assessment has to do with the way in which environmental assessment systems and practices have evolved. Since the early 1970s, much of the development and evolution of environmental assessment has occurred within three silos: project proponents operating in the silo of EIA to get approvals for their projects; scientists operating in the silo of cumulative effects and regional studies to understand broader ecosystem and landscape functioning; and land-use planners and environmental managers operating at the strategic level, above the project tier, focused on futures and broader environmental planning and management concerns. Each of these silos is valuable; however, none of them can be expected to deliver the benefits of the others. It is suggested here that the real, substantive benefits of impact assessment should be measured in terms of its contribution to environmental management and to better decisions about the sustainability of development. This cannot be realized within any individual silo of environmental assessment but only through a more regional, integrative approach. This is not to say that another layer of environmental assessment is required; rather, there is a need to integrate current knowledge and understanding of environmental assessment systems and practices and to do so at the regional and strategic tier if any of the silos is to have meaning.

Towards a Regional Strategic Assessment System

Recent initiatives of the Canadian Council of Ministers of the Environment, Environmental Assessment Task Group (EATG), have focused on the advancement of Regional Strategic Environmental Assessment (R-SEA). In February 2008, the EATG commissioned a report 'Strengthening the foundation for Regional Strategic

Environmental Assessment in Canada' (Noble and Harriman 2008). Based on merging the principles of regional CEA and SEA, Noble and Harriman define R-SEA as:

> *a process designed to systematically assess the potential environmental effects, including cumulative effects, of alternative strategic initiatives, policies, plans, or programs for a particular region.*

The underlying notion of R-SEA is a more integrative, regionally based assessment framework, operating above the project tier and ensuring that knowledge and understanding about the cumulative effects of future development possibilities trickle down to inform development impact assessment and decision-making. The overall objective of R-SEA is to inform the preparation of a preferred development strategy and environmental management framework for a region. In this regard, R-SEA is intended to:

- improve the management of cumulative environmental effects;
- increase the effectiveness of project-level environmental impact assessment;
- identify preferred directions, strategies, and priorities for the future management and development of a region.

It is generally acknowledged that 'good' CEA demands a regional approach (Harriman and Noble 2008; Duinker and Greig 2006; Creasey 2002; Hegmann et al. 1999), but at the same time CEA is also a required component of project-based EIA. Increasing the scale of EIA from the project to the region, however, is beyond the ability of project proponents and is itself insufficient to ensure an understanding of the sources or drivers of cumulative change or to proactively manage development. R-SEA, on the other hand, represents a different way of approaching the interrelationships between environment and development (Noble 2008)—one that enables CEA to occur beyond the constraints of project-based thinking, focusing on desirable futures and outcomes but at the same time ensuring that strategic-level decisions are useful and relevant to informing and influencing the nature of project-based development.

In other words, R-SEA is about identifying and assessing future possibilities; it is focused on informing development decisions for a region, including the nature and types of development that can or should take place based on regional environmental, social, economic, or sustainability goals and objectives and on the potential cumulative environmental effects of alternative options. In this way, R-SEA is a means of identifying preferred development plans, conservation planning initiatives, or resource management programs, which include spatial and temporal considerations of cumulative change, and thus provides direction for project EIA and for decisions about development. Having such a framework of environmental assessment in place is critical to ensuring effective regional environmental planning and cumulative effects management and sustainability at the project level (Creasey 2002). A more regional and strategic-based approach to environmental assessment in Canada has the potential to:

- facilitate effects analysis at broader, regional scales and thereby provide the opportunity to consider cumulative environmental effects before project developments take place;

- allow for the consideration of strategic alternatives early in decision-making before irreversible decisions are taken;
- provide a focus for project EIA with regard to how proposed actions fit within the broader context of the region and a benchmark against which the effectiveness of EIA can be evaluated;
- ensure that subsequent development and project-based assessments are the result of desired rather than most likely outcomes;
- provide a framework to facilitate the multi-agency cooperation necessary to manage and monitor regional cumulative environmental change (see Kingsley 1997; Creasey 2002; Fischer 2002; Cooper and Sheate 2004; Noble 2005; Harriman and Noble 2008; Noble and Harriman 2008).

PROSPECTS

Environmental assessment has come a long way since first introduced to Canada more than 35 years ago. There have been continual process improvements to project-based EIA, and new tools and frameworks, including CEA and SEA, have emerged. At the same time, however, it must be acknowledged that EIA in the absence of more strategic and regional processes is inherently constrained in its ability to facilitate decisions about development that are consistent with the broader principles of sustainability. EIA is focused on project developments; the aim of the proponent is to get project approval by mitigating, to the point of acceptability, potentially adverse environmental effects. Further, at the strategic tier of SEA, these higher-order decisions and planning objectives are of little benefit if they do not translate to the regional and project level and inform decisions about the nature and acceptability of development. Each of these processes—EIA, CEA, and SEA—is valuable in its own right, but each is limited in terms of what it can deliver in the absence of a broader, more integrative, more supportive, and more cyclical system of environmental assessment and knowledge translation from the strategic to the regional and the project level.

The knowledge needed to advance the current state-of-the-art of environmental assessment beyond its current limits does exist, but this knowledge is contained within the individual silos of environmental assessment, and the institutional capacity and the will for the types of actions and collaborations necessary for that advancement seem to be lacking. Far more attention has been given to critiquing environmental assessment than to advancing systems and practices and to ensuring substantive outcomes. That said, and as Therivel and Ross (2007) note, these are still the early days of strategic and integrative thinking in environmental assessment, and we can expect considerable advances in the years to come.

REFERENCES

Bailey, J. 1994. 'EIA, management and policy reform: The tyranny of small decisions working well'. Paper presented at the 14th annual meeting of the International Association for Impact Assessment, Quebec City.

Boyden, A. 2007. 'Environmental assessment under threat'. *International Association for Impact Assessment Newsletter* 19 (1): 4.

Cooper, L., and W. Sheate. 2004. 'Integrating cumulative effects assessment into UK strategic planning: Implications of the European Union SEA Directive'. *Impact Assessment and Project Appraisal* 22 (1): 5–16.

Creasey, R. 2002. 'Moving from project-based cumulative effects assessment to regional environmental management'. In A. Kennedy, Ed., *Cumulative Environmental Effects Management: Tools and Approaches*. Calgary: Alberta Society of Professional Biologists.

Dubé, M. 2003. 'Cumulative effect assessment in Canada: A regional framework for aquatic ecosystems'. *Environmental Impact Assessment Review* 23: 723–45.

Duinker, P., and L. Greig. 2006. 'The impotence of cumulative effects assessment in Canada: Ailments and ideas for redeployment'. *Environmental Management* 37 (2): 153–61.

Fischer, T. 2002. *Strategic Environmental Assessment in Transport and Land Use Planning*. London: Earthscan.

Fuggle, R. 2005. 'Have impact assessments passed their "sell by" date?' *International Association for Impact Assessment Newsletter* 16 (1): 1, 6.

Galbraith, L., B. Bradshaw, and M. Rutherford. 2007. 'Towards a supraregulatory approach to environmental assessment in northern Canada'. *Impact Assessment and Project Appraisal* 25 (1): 27–41.

Gibson, R.B. 2002. 'From Wreck Cove to Voisey's Bay: The evolution of federal environmental assessment in Canada'. *Impact Assessment and Project Appraisal* 20 (3): 151–9

Grzybowski and Associates. 2001. 'Regional environmental effects assessment and strategic land use planning in British Columbia'. Report prepared for the Canadian Environmental Assessment Agency Research and Development Program. Hull, QC: Canadian Environmental Assessment Agency.

Harriman, J., and B. Noble. 2008. 'Characterizing project and regional approaches to cumulative effects assessment in Canada'. *Journal of Environmental Assessment Policy and Management* 10 (1): 25–50.

Hegmann, G., et al. 1999. *Cumulative Effects Assessment Practitioners Guide*. Prepared by AXYS Environmental Consulting and CEA Working Group for the Canadian Environmental Assessment Agency. Hull, QC.

Jay, S., et al. 2007. 'Environmental impact assessment: Retrospect and prospect'. *Environmental Impact Assessment Review* 27: 287–300.

Kingsley, L. 1997. *A Guide to Environmental Assessments: Assessing Cumulative Effects*. Hull, QC: Parks Canada, Department of Canadian Heritage.

Noble, B.F. 2005. *Regional Cumulative Effects Assessment: Toward a Strategic Framework*. Research supported by the Canadian Environmental Assessment Agency's Research and Development Program. Ottawa: Canadian Environmental Assessment Agency.

———. 2008. 'Strategic approaches to regional cumulative effects assessment: A case study of the Great Sand Hills, Canada'. *Impact Assessment and Project Appraisal* 26 (2): 78–90.

———. 2009. 'Promise and dismay: The state of strategic environmental assessment systems and practices in Canada'. *Environmental Impact Assessment Review* 29 (1): 66–75.

Noble, B.F., and J. Harriman. 2008. *Strengthening the Foundation for Regional Scale Strategic Environmental Assessment in Canada*. Research report prepared for the Canadian Council of Ministers of the Environment Environmental Assessment Task Group under contract agreement. Ottawa: Canadian Council of Ministers of the Environment.

Noble, B.F., and K. Storey. 2005. 'Toward increasing the utility of follow-up in Canadian EIA'. *Environmental Impact Assessment Review* 25 (2): 163–80.

Ross, W. 1994. 'Assessing cumulative environmental effects: Both impossible and essential'. In A. Kennedy, Ed., *Cumulative Effects Assessment in Canada: From Concept to Practice*. Papers from the 15th Symposium held by the Alberta Society of Professional Biologists, Calgary.

Spaling, H., et al. 2000. 'Managing regional cumulative effects of oil sands development in Alberta, Canada. *Journal of Environmental Assessment Policy and Management* 2 (4): 501–28.

Storey, K. 1986. 'From prediction to management: Increasing the effectiveness of social impact assessment'. In H. Becker and A. Porter, Eds, *Impact Assessment Today*, v. II. Utrecht: Uitgeverij Jan van Arkel.

Therivel, R., and W. Ross. 2007. 'Cumulative effects assessment: Does scale matter?' *Environmental Impact Assessment Review* 27: 365–85.

APPENDIX I

Alternative Means: An Exercise in Project Assessment

INSTRUCTIONS

The purpose of this exercise is to provide course participants with an opportunity to apply various EIA methods, techniques, and procedures discussed in this guide to a development problem that mimics a real-world situation. Students should work in groups (consulting teams) to prepare a preliminary scoping and assessment report, including a project assessment matrix, criteria and a protocol for determining impact significance, suitable techniques for predicting impacts, and management and impact monitoring mechanisms. The exercise is designed to be applied to the 'local environment' within which the course is being delivered; thus, it is necessary that the students be provided with copies of topographic and political maps of the region.

HOUSEHOLD AND LIGHT INDUSTRIAL WASTE SANITARY LANDFILL PROJECT

Background

Waste Management Inc. recently filed an application for the development and operation of a regional sanitary landfill site for household and light industrial waste. You have been contracted by the company to prepare a preliminary scoping and assessment report in preparation for a full environmental assessment of the proposed facility.

The proposed landfill project will consist of three main components:

- construction of a site access road;
- excavation and operation of an open-pit waste disposal site;
- project and site decommissioning.

Household and light industrial waste from regional communities will be shipped to the disposal site via highway transport. The construction of a single-lane, gravel-surfaced site access road capable of withstanding the load of the heavy equipment used to ship the waste material is required. Site development will require the excavation of a large open pit, to be lined with an impermeable polyurethane sheet to avoid hydraulic contact with groundwater. The landfill pit design is depicted in Figure A.1 below.

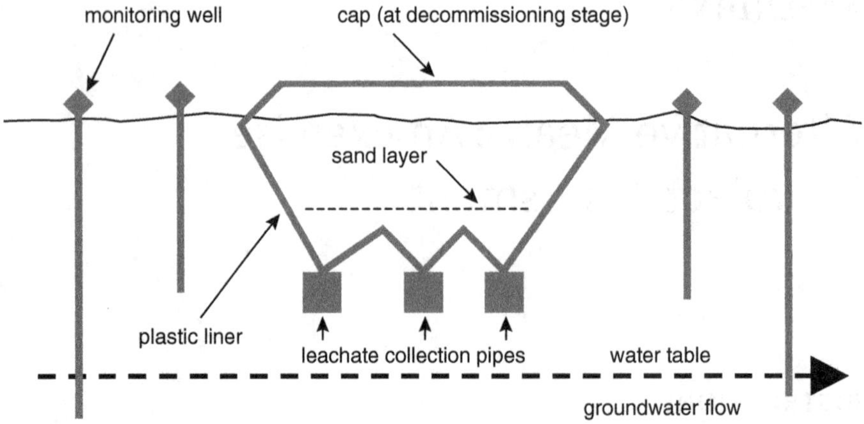

Figure A.1 Landfill project engineering Design.

Total employment during construction is estimated to peak at 125 persons during the first three months of the project. It is anticipated that on average, 70 per cent of these jobs will be filled by workers from the region, primarily for labour-intensive work associated with pit excavation and road construction. Analysis of the local labour force suggests that about 30 per cent of the peak labour force requirements will be met by local residents. Construction is expected to last for eight months.

The design life of the facility is approximately 12 years, during which time the permanent on-site workforce will be about 25 persons, including site managers, site maintenance, and security. At the end of this period, the disposal site will be decommissioned. This will involve covering the waste with compacted clay, a low-permeability layer of synthetic material, a drainage layer of stone and gravel, and a vegetated top. Monitoring wells will remain intact and operable.

The total cost of the project is estimated at $26.5 million, including the access road and landfill construction and site decommissioning.

Report Requirements

As the hired consultant, you are asked to deliver a report detailing the following:

Location Analysis

Construct a general 'screening checklist' of location criteria for a sanitary landfill project (e.g., 50 metres from a waterway, minimum of five kilometres from a residential area, etc.), and identify a potential project location on your map sheet.

Project and Baseline Description

A description of the project actions and current environmental baseline is required. This will consist of three components:

1. A list of project actions associated with the construction and operation of the sanitary landfill, including, for example, site clearing, pit construction, and waste transport.

2. Description of the human and biophysical baseline environment of the locality, including population, labour force statistics, and existing environmental quality and stress.
3. Identification of the likely affected and interested publics for consultation.

VEC Identification
Identify potentially affected human and biophysical VECs, and delimit an appropriate spatial boundary for the assessment.

Impact Identification
The proponent requests that two impact identification components be included in the preliminary report:

- A simple network visually identifying the linkages and order of impacts associated with project actions and affected VECs.
- An impact matrix identifying and characterizing the potential impacts on the affected VECs. The nature of the impact matrix is at the discretion of the consultant but must include: (1) project actions and affected environmental components; (2) some indication (e.g., scale or weighting) of the potential impact significance, which should be clearly explained; (3) a one-sentence statement of each of the identified impacts.

Techniques for Impact Prediction
Once all impacts are identified, the proponent has requested that the consultant identify four of the most significant environmental impacts, of which at least one must concern the biophysical environment and one the human environment and one must be a positive impact. For each impact, provide a maximum two-page description of the predictive technique(s) the proponent might use for impact prediction. A brief assessment of the strengths and limitations of the predictive technique should also be included.

Impact Management Measures
For each of the above impacts, recommendations should be made as to appropriate and feasible impact management and mitigative measures. Impact management measures for each impact should be detailed on one page.

Monitoring Components
A monitoring program is required in order to measure actual environmental change and to evaluate the effectiveness of impact management measures. For each of the four significant impacts:

- identify appropriate impact indicators for monitoring;
- identify the type of data that should be collected for each of these indicators and general location(s) where the data are to be collected.

Additional information and references, including a project location map, should be appended to the consulting report.

APPENDIX II

Alternatives to: An Exercise in Strategic Assessment

INSTRUCTIONS

The purpose of this exercise is to provide course participants with an opportunity to apply various SEA methods, techniques, and procedures discussed in this guide to a policy and planning problem that mimics a real-world situation. Students should work in groups (consulting teams) to prepare a SEA report, including an assessment matrix, criteria and a protocol for determining significance, and identification of the preferred strategic direction. The exercise is designed to be applied to the 'regional or national environment' within which the course is being delivered.

ELECTRICITY SUPPLY SOURCE ALTERNATIVES

Background

The demand for electricity within the region is projected to increase at an annual average rate of 1.8 per cent over the next 20 years. The current electricity supply system is expected to reach its capacity within the next two years, and a strategy is required now in order to plan for the future and appropriately address electricity demand projections.

A recent workshop involving government and various industry representatives identified a number of technologically feasible alternatives for addressing the projected increase in electricity demand, including:

Option 1: An intensive energy conservation and demand reduction program
Option 2: The development of one new coal-fired generation station
Option 3: An increase in renewable energy—namely, wind power
Option 4: An increase in hydroelectric generation capacity
Option 5: The development of one new nuclear reactor

Report Requirements

You have been contracted by the proponent—the provincial government—to provide an assessment of supply alternatives based on your expert judgment and consultation with other experts. You are asked to deliver a report detailing the following:

Assessment Factors

A list of assessment factors and criteria is required in order to assess the implications of each proposed alternative. Included on this list should be economic factors and objectives (e.g., minimize electricity cost to consumers), social factors (e.g., ensure equitable access and distribution), and environmental factors (e.g., minimize atmospheric emissions). Criteria should be limited to the seven to nine criteria that are deemed most important.

Criteria Weighting

The assessment factors are to be weighted in order to ascertain their relative importance in the decision and assessment process. Include in the report a description of the weighting procedure and a table of final weights for each criterion. Given that this is a preliminary assessment, the proponent requests that weights be derived on the basis of your expert judgment and experience.

Alternatives Assessment

Each alternative is to be assessed against the proposed criteria. Include in the report a description of the methods used for assessment—namely, those that rely on your expert judgment and experience, as well as the necessary tables or impact matrices indicating the impacts or preferences for each alternative on the basis of the individual criterion.

Identification of Strategic Option(s)

Identify, based on the above assessment, an overall ranking of electricity supply options.

Sensitivity of Decision

Evaluate the sensitivity of the rankings and assessment outcomes by examining two critical scenarios of particular importance to the preferred electricity supply option. First, examine the implications of the development of a stringent emissions policy by increasing the weights assigned to any environmental factors and criteria by 50 per cent. Second, examine the implications of an increase in the cost of electrical generation by assigning a 50 per cent increase to the economic factors and criteria. Discuss the outcomes or implications of these changes with regard to the ranking of supply options.

Additional information and references, including a project location map, should be appended to the consulting report.

Glossary of Key Terms

accuracy: The amount of bias applied to the system-wide impact predictions.

acid mine drainage: Drainage from surface mining, deep mining, or coal refuse piles, usually highly acidic with large concentrations of dissolved metals.

active publics: Publics that affect assessment decisions.

activity information: The effects possibly generated as a result of a project.

ad hoc approaches: Applying EIA findings of past common projects to predict or understand proposed or existing projects.

adaptive environmental assessment and management (AEAM): A simulative modelling approach that quantifies and defines resources and environmental problems.

additive effects: Individual minor actions that may be separate or related and that have a significant overall impact when combined.

adverse effects: Direct project effects that have potential to cause long-term, irreversible, undesirable environmental damage or change.

alternative means: Different ways of carrying out a proposed project—typically alternative locations, timing of activities, or engineering designs.

alternatives to: Under the Canadian Environmental Assessment Act, 'alternatives to' refers to different ways of addressing the problem at hand or meeting the proposed project objectives; renewable energy, for example, would be considered an 'alternative to' a proposed coal-fired generating station.

ambient environmental quality monitoring: A pre-project assessment of the surrounding environment inclusive of biophysical and socio-economic factors; collected information is used as a baseline in comparing a project's environment during development, operation, and post-operation against unaffected control sites in order to monitor the impact of a project.

ambitious bounding: Selecting a project boundary that is too large, often leading to the proposed project seeming insignificant.

amplifying effects: Incremental effect additions whereby each increment has a larger effect than the one preceding it.

analysis scale: The scale applied to examine VECs and impacts across space.

antagonistic effects: Individual adverse effects that have potential to partially cancel each other out when combined.

assimilative capacity: The ability for an environment to incorporate pollutants into its system without adverse effects to its natural state.

auditing: An objective examination or comparison of observations with predetermined criteria.

backcasting: The process of working backwards from a particular future condition judged to be desirable and then determining the feasibility of achieving that condition given project actions and changing environmental conditions.

balance model: Model designed to identify inputs and outputs for specified environmental components; these models are commonly used to predict change in environmental phenomena.

baseline condition: A description of the pre-project environment, which is inclusive of the cumulative effects of previous activities and the future environment in the absence of the proposed project.

baseline study: Description of the biophysical and socio-economic state of the environment at a given time that can later be used for the comparison of environmental change through time.

biophysical environment: Consists of air, water, land, rocks, minerals, plants, animals, and the interacting components of ecosystems.

buffer zone: An area of undisturbed environment; a commonly used mitigation practice.

Canadian Environmental Assessment Act: The legal basis for federal EIA in Canada; sets out responsibilities and procedures for EIA of projects that involve federal authorities; introduced in Parliament in 1992, proclaimed in 1995, and most recently revised in 2003.

Canadian Environmental Assessment Agency: The federal agency created in 1994, replacing the Federal Environmental Assessment Review Office, to oversee Canadian federal EIA and implementation of the Canadian Environmental Assessment Act.

case-by-case screening: Evaluating project characteristics against a checklist of regulations and guidelines.

checklists: Method used to create a comprehensive list of effects or indicators of environmental impacts that a project might generate.

chemical antagonism: A reaction between two effects, such as in a chemical reaction, whereby the severity of the combined effect is reduced.

class screening EIA: An assessment method used to streamline projects deemed unlikely to cause adverse environmental effect or that are considered routine undertakings.

closed scoping: EIAs with content and scope predetermined by law; modifications can only be made through closed consultation between the proponent and the responsible authority or regulatory agency.

compensation: The measures taken by the proponent to make up for adverse environmental impacts of a project that exist after mitigation measures have been implemented.

compliance monitoring: Monitoring to ensure that all project regulations, agreements, laws, and specific guidelines have been adhered to.

component interaction matrices: Applied to improve understanding of indirect impacts from projects; the matrices identify first-, second-, and high-order impacts and illustrate the dependencies between environmental components.

comprehensive study EIA: An EIA assessment applied on large-scale, complex, environmentally sensitive projects that have a high risk of causing adverse environmental effects.

confirmatory analysis: Used to test for uncertainty in impact predictive techniques and to ensure similar predictive outcomes from different types of techniques.

consistency: A measure of the quality of an expert's judgment, including a test of the degree of randomness in a set of assessment scores or judgments.

continuous effects: Effects that are ongoing over space or time.

control site: A reference point where the environment is not affected by the project; used to monitor the nature and extent of project-induced environmental change in areas that are affected by the project.

cost-benefit analysis: An assessment method that expresses project impacts based in monetary terms.

cumulative effects assessment (CEA): The assessment of changes to the environment that are caused by actions in combination with other past, present, and future actions; changes may be linear, amplifying or exponential, discontinuous, or structural surprises.

cumulative effects monitoring: Non-site-specific monitoring that emphasizes the accumulated effects of multiple developments within a particular region.

decision-point audit: Examination of role and effectiveness of the EIS based on whether the project is allowed to proceed and under what conditions.

Delphi technique: An iterative survey-type questionnaire that solicits the advice of a group of experts, provides feedback to all participants on the statistical summaries of the responses, and gives each expert an opportunity to revise her/his judgments.

descriptive checklist: Type of EIA checklist that gives guidance on how to assess impacts and what information is required.

deterministic model: A model dependent on fixed relationships between environmental components.

direct effects: First-order impacts that have a particular social value.

discontinuous effects: Effects that occur in an incremental manner and go unnoticed until some threshold is reached.

dispositional antagonism: A type of antagonistic effect in which one effect affects the uptake or transport of another effect, such as ethanol enhancing mercury elimination in mammals.

draft EIS audit: Review of the project EIS according to its terms of reference.

duty to consult: A formal, legal obligation in Canada for governments to consult with Aboriginal people in cases where Aboriginal rights, claims, or titles are known and may be affected by a development or decision, even in cases where those rights, claims, or titles have not yet been proven in court.

early warning indicators: Biological or non-biological indicators that can be measured to detect the possibility of adverse stress on VECs before they are adversely affected.

effects-based CEA: Measures multiple environmental responses to stressors; the results are then compared to some control point to determine the actual measure of cumulative change.

environment: Refers to all environments, including biophysical, social, and economic components.

Environmental Assessment Review Process: The first Canadian federal EIA process, formally introduced in 1973 by guidelines order.

environmental baseline: The present and likely future state of the environment without the proposed project or activity.

environmental change: Measurable change in an environmental parameter over time.

environmental effect: The difference in the condition of an environmental parameter with as opposed to without a proposed development activity.

environmental impact assessment (EIA): A systematic process designed to identify, predict, and propose management measures concerning the possible implications that a proposed project's actions may have for the environment.

environmental impact statement (EIS): The formal documentation produced from the EIA process that provides a non-technical summary of major findings, statement of assessment purpose and need, detailed description of the proposed action, impacts, alternatives, and mitigation measures.

environmental impacts: Environmental effects that have an estimated societal value placed on them.

environmental management system: A voluntary industry-based management system that controls the activities of products and processes that could cause adverse environmental impacts; management systems are a cyclical process of continual improvement in which industries and firms are constantly reviewing and revising their way of doing business to protect the environment.

environmental preview report: An early report that is used to determine whether or not an EIA is needed and the level of assessment required; it outlines the project, potential environmental effects, alternatives, and management measures.

environmental protection plans: Mandatory management plans that result from project-based EIAs; these management plans are tailored to the project as a result of the identification of key impacts and issues and management measures through the EIA process.

environmental site assessment: An environmental study for the purposes of determining the nature and extent of contamination of a local site and prescribing cleanup and reclamation plans.

environmental systems: Environmental components functioning together as a unit.

exclusion list: A screening mechanism listing projects that would be subject to an EIA unless they were *included* in the list; generally, projects excluded involve issues of national defence and emergency or are projects routine in nature.

experimental monitoring: Research into environmental systems and their impacts for the purpose of gathering information and knowledge and testing hypotheses.

ex-post evaluation: Taking action and making decisions based on the result of structure, analysis, and appraisal of information concerning project impacts.

Federal Environmental Assessment Review Office (FEARO): The federal agency created to oversee the implementation of the Federal Environmental Assessment Review Process.

fixed-point scoring: A VEC weighting approach in which a fixed number of values is distributed among all affected environmental components; the higher the point score, the more important the environmental component.

fly-in fly-out: Projects located in remote areas with high amounts of air traffic to and from the site; typically associated with remote mining projects where temporary project work camps are constructed to house workers, who commute by charter air service on a basis, for example, of two weeks at the work site and two weeks off during the lifespan of the project.

functional antagonism: A type of antagonistic effect whereby one effect counterbalances an organism's physiological response to a second effect.

functional scale: Scale relationship based on how different environmental components function across space.

Gaussian dispersion model: A model devised for predicting point-source atmospheric pollution.

Geographic Information System (GIS): A system of computer hardware and software for working with spatially integrated and geo-referenced data.

gradient-to-background monitoring: Monitoring system that measures the effects caused by the impact source at an increasing distance from the impact origin to the point of background assimilation; an 'artificial' control point is established.

gravity model: A deterministic model used to predict population flow or spatial interaction; the model is dependent on a fixed and inverse relationship between mass (population) and distance.

holistic approach: Examining environmental components on the basis of their position and functioning within the context of the broader environmental system.

human environment: Aspects of the environment that are non-biophysical components; also referred to as the socio-economic or cultural environment.

impact avoidance: A form of impact management whereby impacts are avoided at the outset by way of alternative project designs, timing, or location rather than managed or mitigated after they occur.

impact benefit agreement: Legal agreement between a proponent and a community or group that will potentially be affected by a project; generally applied to ensure that the resources for maximizing the benefits associated with the development are fully capitalized on.

impact mitigation: Minimizing adverse environmental change associated with a project by implementing environmentally sound construction, operating, scheduling, and management principles and practices within project design.

impact significance: Reflects the degree of importance of an impact and is based on the characteristics of the impact, receiving environment, and societal values.

implementation audit: An evaluation of whether or not the recommendations presented in a project's EIS were actually put into practice.

inactive publics: Publics not normally involved in the environmental planning, decisions, or project issues.

inclusion list: A screening method listing projects that have mandatory or discretionary requirements for an EIA.

incremental effects: Marginal changes in the condition of an environmental component caused by project actions.

initial environmental examination: Information prepared to establish whether or not an EIA is needed and what level of EIA should be implemented.

input evaluation: Follow-up or auditing of SEA procedures and requirements, such as purpose and objectives, data quality, and linkages to policy, planning, or program initiatives.

inspection monitoring: Site-specific monitoring with on-site visits to ensure compliance with procedures and safety standards.

intention survey: A survey that attempts to collect the judgment of as many people as possible and record their responses regarding what they intend to do or how they might react, given certain circumstances or situations.

interaction matrix: A type of EIA matrix based on the multiplicative properties of simple matrices to generate a quantitative impact of the proposed project on interacting environmental components.

ISO 14001: Internationally recognized industry performance standard for environmental management system certification.

Keynesian multiplier: A basic economic multiplier that notes that an injection of money into a local economy will increase at the local level by some multiple of that initial injection.

Leopold matrix: An EIA matrix for identifying first-order project-environment interactions, consisting of a grid of 100 possible project actions along a horizontal axis and 88 environmental considerations along a vertical axis.

life-cycle assessment: The 'cradle-to-grave' assessment of projects from their inception and start-up to post-operation.

linear additive effects: Incremental additions to or from a fixed environment where each additional increment has the same effect.

list-based screening: A checklist of projects that may or may not require an EIA.

Mackenzie Valley Environmental Impact Review Board: A valley-wide public board created as part of the Mackenzie Valley Resource Management Act to undertake EIAs and panel reviews under the jurisdiction of the Act.

Mackenzie Valley Resource Management Act: An Act implemented by the federal government to give decision-making authority to northerners concerning environment and resource development activities within the Mackenzie Valley region of the Northwest Territories; proclaimed in 1998, the Act governs EIA in the region.

magnitude: The size or the degree of a predicted impact; not to be equated with significance.

magnitude matrices: Matrices that attempt to identify impacts and summarize impact importance, time frame, and magnitude.

matrices: Management and assessment tools used for identifying project impacts that typically consist of a two-dimensional checklist that places project actions on one axis and environmental components on the other.

maximum allowable effects level: An approach to impact prediction based on specifying certain desired limits or thresholds that a certain impact is not to exceed.

mechanistic model: A type of model based on mathematical equations or flow diagrams that describe cause–effect relationships in a project environment.

mediation EIA: An approach to EIA in which an independent neutral mediator helps to resolve conflict between different interest groups, stakeholders, and proponents throughout the assessment.

methods: The various aspects of an assessment, including organization, identification of impacts, and collection and classification of data.

model class screenings: Provide a generic assessment of all projects within a certain class in which information contained in a 'model' report is prepared for individual projects and adapted for location- or project-specific information.

models: Box-and-arrow or mathematical equations used to simplify real-world environmental systems.

monitoring: A systematic process of data collection or observations used to identify the cause and nature of environmental change.

monitoring for knowledge: The monitoring used after impacts occur; data are collected and used for future impact prediction and project management.

monitoring for management: Tracking and evaluating changes in a range of environmental, economic, and social variables; usually associated with high-profile projects with uncertain outcomes and with the potential for significant adverse outcomes.

monitoring of agreements: Monitoring and auditing of agreements between project proponents and affected groups to ensure compliance.

Monte Carlo analysis: A means of statistical evaluation of mathematical functions using random samples.

multiplier: A quantitative expression of some initial exogenous change and the expected additional effects caused by interdependences with an endogenous linkage system.

National Environmental Policy Act (NEPA): The US legislation of 1969 that required certain development project proponents to demonstrate that their projects would not cause adverse environmental effects; the beginning of formal EIA.

networks: An EIA method used to identify direct and indirect impacts that may be triggered by early project activities.

non-point-source stress: Environmental stress from diffuse sources that cannot be traced back to a particular project origin, such as runoff from urban areas or pollution introduced to streams from groundwater.

Nunavut Impact Review Board: Established under the Nunavut Land Claims Agreement and the primary authority responsible for EIA in the land claims area.

Nunavut Land Claims Agreement: Canada's largest land claims settlement and land claims–based EIA process; signed in 1993, giving the Inuit self-governing authority and leading to the establishment of a new territory, Nunavut, in 1999.

off-site impacts: Impacts that occur at a distance or removed from the recognized project area.

on-site impacts: Impacts that occur directly in the immediate project area.

open scoping: A transparent scoping process in which the content and scope of the assessment are determined through consultation with various interests groups and public stakeholders.

output evaluation: Monitoring or auditing SEA effectiveness on the basis of the output of the SEA process—namely, whether the process informed policy and whether the intended outcomes were actually realized.

paired comparisons: An approach to determining the relative importance of impacts in a hierarchy in ratio form whereby the decision-maker considers trade-offs one at a time for each pair of VECs.

performance audit: An assessment of a proponent's capability to respond to environmental incidents and of its management performance.

Peterson matrix: A multiplicative EIA matrix consisting of project impacts and causal factors, resultant impacts on the human environment, and relative importance of those human components used to derive an overall project impact score.

phenomenon scale: The scale used to determine the spatial extent within which certain environmental components and VECs operate and function.

plan: A defined strategy or proposed design to carry out a particular course of action or several actions and various options and means to implement those actions.

point-source stress: Environmental stress that can be traced back to a particular project origin, such as discharge or emissions from a development activity.

policy: A guiding intent, set of defined goals, objectives, and priorities, either actual or proposed.

policy-based SEA: Strategic environmental assessment applied to policies that have no explicit 'on-the-ground' dimension, such as fiscal policies or national energy policies; also referred to as indirect SEA.

precautionary principle: When information is incomplete but there is threat of an adverse effect, the lack of full certainty should not be used as a reason to preclude or postpone actions to prevent harm.

precision: The exactness of impact prediction.

predictive technique audit: A type of environmental audit in which a project's predicted effects are compared to the actual effects.

probability analysis: An analysis that uses quantified probability to classify the likelihood of an impact occurring and under what environmental conditions.

process evaluation: Monitoring and auditing SEA based on the actual SEA process, including methods, techniques, openness, and frameworks.

program: A schedule of proposed commitments or activities to be implemented within or by a particular sector, plan, or area of policy.

programmatic environmental assessment: Under the US NEPA, the application of environmental assessment to multiple projects or to programs of development.

programmed-text checklist: A type of EIA checklist consisting of a series of filter questions for project screening and impact identification; useful for standard or routine projects.

project evaluation monitoring: Performance auditing or productivity measurement, a monitoring program concerned with a project's performance and ability to reach specified goals and objectives.

project impact audit: A type of auditing that focuses on determining whether the actual project impacts were predicted in the EIS.

questionnaire checklist: An EIA screening method consisting of a set of questions that must be answered when considering the potential effects of a project.

rating: An approach to VEC or impact weighting whereby the importance or significance of each is indicated on a numerical scale ranging from, for example, 1 to 5; no direct decision trade-offs are involved.

receptor antagonism: A type of antagonistic effect in which one effect blocks the other, such as a toxicant binding to a receptor and blocking the effects of a second toxicant.

receptor information: Information pertaining to the processes resulting from project-induced effects, such as habitat fragmentation, that is important to consider when characterizing the environmental setting.

rectifying impacts: An approach to impact management based on restoring environmental quality, rehabilitating certain environmental features, or restoring environmental components to certain degrees following an environmental impact.

regional-based SEA: An approach to SEA concerned with regional-based environmental planning or development and assessing the impacts of area-specific plans and program initiatives.

regional study–based CEA: Cumulative effects assessment applied over a broader spatial scale to assess a wide range of impacts in a specific region or area.

regulatory permit monitoring: Site-specific monitoring that includes regular documentation of requirements necessary for permit renewal or maintenance.

relative significance: A measurement of the relative importance of one affected environmental component over another.

replacement class screening: A generic assessment of all projects within a class, but no location- or project-specific information is required, and thus a screening report is not necessary for each individual project.

residual effects: Effects that remain after all management and mitigation measures have been implemented.

restrictive bounding: An approach to project scoping in which the spatial area identified for consideration in the impact assessment is perhaps too small for a complete understanding of total environmental effects and within which the impact of the proposed project may be inflated because of the small area under consideration.

review panel EIA: A level of EIA that is applied to projects with uncertain or potentially significant effects or if public and stakeholder concern warrants an independent review panel.

risk: The possibility that an undesired outcome may result from an uncertain situation.

risk assessment: Using collected information to identify potential risks.

sanitary landfill: A type of waste disposal site where solid waste is contained within an impermeable barrier within the earth's surface and covered.

scenario analysis: An approach used in EIA and SEA to identify hypothetical actions or situations and potential outcomes.

scoping: An early component of the EIA process used to identify important issues and parameters that should be included in the assessment.

screening: The selection process used to determine which projects need to undergo an EIA and to what extent.

screening EIA: A type of project assessment that is an extension of the basic screening process in which anticipated environmental effects are documented and the need for additional project modification or further assessment is determined.

secondary impacts effects: An impact resulting from a direct impact.

sector-based SEA: A type of SEA based on initiatives, plans, and programs that are specific to certain industrial sectors.

sensitivity analysis: Examination of the sensitivity of an impact prediction to minor differences in input data, environmental parameters, and assumptions.

socio-economic monitoring: A type of monitoring that looks specifically at socio-economic parameters in the review of a project area.

Sorensen network: A hybrid of an EIA matrix and simple network that identifies direct impacts triggered by project actions and cause-effect relationships.

spatial scale: The actual geographic scale used to define the extent of a project EIA.

statistical model: A type of model used to test relationships between variables and to extrapolate data.

statistical significance: The determination, based on confidence intervals and probabilistic data, as to whether a particular outcome or prediction is significant based on theoretical and empirical findings.

stochastic model: A type of mechanistic model that is probabilistic in nature or gives an indication of the probability of an event occurring within specified spatial and temporal scales.

strategic environmental assessment (SEA): The environmental assessment of initiatives, policies, plans, and programs and their alternatives.

stressor-based CEA: Assessment that predicts cumulative effects associated with a particular agent of change.

structural surprises: Cumulative effects that occur in regions with multiple developments; they are the least understood and the most difficult cumulative effects to assess.

sustainability assessment: Broadly defined as a process by which the implications of an initiative on sustainability are evaluated to help decision-makers decide what actions to take or not to take to make society more sustainable.

synergistic effects: When the total effects are greater than the sum of the separate, individual effects.

systems diagrams: Models based on box-and-arrow diagrams that consist of environmental components linked by arrows indicative of the nature of energy flow or interaction between them.

techniques: Ways of providing and analyzing data in EIA.

threshold-based prediction: Basing impact predictions on prior experiences using approaches such as maximum allowable effects levels whereby an impact is capped and not to exceed a certain threshold or level of change.

threshold-based screening: A screening process whereby proposed developments are placed in categories and thresholds are set for each type of development, such as project size, level of emissions generated, or area affected.

threshold of concern checklist: A type of EIA checklist that lists environmental components that might be affected by the project actions, specific criteria for each component, and thresholds against which the project actions can be assessed.

traditional ecological knowledge: Local or Aboriginal knowledge acquired from experience, culture, or interaction with land and resources over time.

valued ecosystem components (VECs): Components of the human and physical environment that are considered important and therefore require evaluation within EIA.

VEC indicators: Provide a measure of qualitative or quantitative magnitude for an environmental impact and might include, for example, specific parameters of air quality, water quality, or employment rates; allow decision-makers to gauge environmental change efficiently.

VEC objectives: The specific parameters, guidelines, or standards set for potentially affected VECs.

weighted impact interaction matrices: An EIA matrix method whereby impacts are multiplied by the relative importance of the affected environmental components and secondary impacts are explicitly incorporated.

weighted magnitude matrices: An EIA matrix method in which degrees of importance, representing the potential impacts of a particular project action on an environmental component, are assigned to the affected environmental components and then multiplied by project impacts.

Index